# What people are saying about

# The Music of the Divine Spheres

This book is a necessary and fascinati
science and spirituality critical to ou
hidden in plain sight, so too does it
fresh perspective from one who has approached his subject from outside the box of conventional orthodoxy and thus reveals it to seekers of spiritual truth.

Drawing upon clues and information brought forth from ancient knowledge and his own research, Milovanov ties together science and spirituality in a way that underscores and illustrates the divine truth that all things are intimately interconnected: music, dimensions, the ancient pyramids, feelings, sacred geometry, emotions, light, color and sound all are revealed to be eternal aspects of Nature's Way of Life, and as such can become beacons for an awakening humanity ready to graduate to the next level of human evolution.

*The Music of the Divine Spheres* is a further exploration and survey of hidden knowledge and wisdom that is required reading for students and adepts to incorporate into their understanding of consciousness and internal transformation.

**Phil Lang**, authorized teacher of Drunvalo Melchizedek's Awakening the Illuminated Heart® workshop

This is a provocative book stretching the connection between traditional and existential thought.

It is challenging and engaging as is, and in a similar way as, the work of Carl Jung and the Collective Unconscious.

There is much to be "explained" albeit in different ways and perspectives.

This book by Alexander Milovanov does just that—stretches how and why to look at what we traditionally accept as more fact in different perspectives of creative ingenuity.

There is much to be offered from the sciences and engineering, traditional consultancy and thought and psychodynamic theory and modeling and new creative adventuring to alternative perspectives. This, for many, may be a fascinating read that provokes creative thinking and pushes the frontiers of science via discussion of conscious relative to unconscious thinking.

**Thomas Olson**, Professor, University of Southern California, Los Angeles, CA

# The Music of the Divine Spheres

The rediscovered ancient knowledge of human consciousness, sacred geometry, and the Egyptian pyramids that can change your life

# Also by the Author

1. A. Milovanov, *The Music of the Divine Spheres* (in Russian), Belovodie, 2018. ISBN 978-5-93454-245-1.
2. A. Milovanov, *The Civilization of Lost Connection* (in Russian), Belovodie, 2019. ISBN 978-5-93454-256-7.

# The Music of the Divine Spheres

The rediscovered ancient knowledge of human consciousness, sacred geometry, and the Egyptian pyramids that can change your life

Alexander Milovanov

BOOKS

Winchester, UK
Washington, USA

JOHN HUNT PUBLISHING

First published by O-Books, 2023
O-Books is an imprint of John Hunt Publishing Ltd., 3 East St., Alresford,
Hampshire SO24 9EE, UK
office@jhpbooks.com
www.johnhuntpublishing.com
www.o-books.com

For distributor details and how to order please visit the 'Ordering' section on our website.

Text copyright: Alexander Milovanov 2022

ISBN: 978 1 80341 364 8
978 1 80341 365 5 (ebook)
Library of Congress Control Number: 2022945974

A CIP catalogue record for this book is available from the British Library.

Design: Lapiz Digital Services

UK: Printed and bound by CPI Group (UK) Ltd, Croydon, CR0 4YY
Printed in North America by CPI GPS partners

The author of this book does not dispense medical advice or
prescribe the use of any technique as a form of treatment for
physical, emotional, or medical problems without the advice of a
physician, either directly or indirectly. The intent of the author
is only to offer information of a general nature to help you in
your quest for emotional and spiritual well-being. In the event
you use any of the information in this book for yourself, which is
your constitutional right, the author and the publisher assume no
responsibility for your actions.

We operate a distinctive and ethical publishing philosophy in
all areas of our business, from our global network of authors to
production and worldwide distribution.

# Contents

# Preface

This book rediscovers the forgotten sacred knowledge that the ancient Egyptian priests used to activate initiated disciples into their higher levels of consciousness. You can use this knowledge for the same purpose. The higher the level of our consciousness that is activated, the higher-level decisions we are able to make in our life. Most importantly, you will learn how our consciousness is constructed and how you can accelerate the formation of your new levels of consciousness to make your life more successful and fulfilling. *The Music of Divine Spheres* lays out this knowledge in a simple, interesting, and visually compelling form that any person can easily understand. Things created by nature are so perfect that they can be described and visualized using little to no mathematical formulas. That's why the book contains many pictures and drawings.

Alexander Milovanov

# Foreword

In the Infinite, there is no time, no place, no divisions, no sense of "us" and "them." Infinite consciousness is the only truth. The heart only knows Unity and Oneness and is thus able to take us beyond the illusory walls of limited perception.

The right brain is our connection to the All That Is; it is holistic, intuitive, experiential, and knows only momentary time, the eternal now. It is able to intuit the Oneness, like a 4- or 5-year-old, looking up at the night sky. The left brain sees dots and does not connect them. It is logical, and is locked into linear time, so it is never in the present moment.

We need both, however, and we need them harmoniously working together. The mind looks out at the reality and sees separation; it does not therefore believe in Oneness. It must be shown unity in a series of logical steps; and Alexander Milovanov has done a masterful job of showing us the unity of all life through the universal language of sacred geometry.

When the mind (left brain) truly sees unity, the corpus callosom (neural fibers connecting the two hemispheres) opens up, communication happens, integration takes place, a relaxation occurs and we become whole again. Because mind is now serving consciousness instead of serving itself, we can replace beliefs with intuitive knowing and learn to follow Spirit. The doorway to the heart has thus been opened.

*The Music of the Divine Spheres* makes a significant contribution to the great awakening that is currently unfolding on a global scale.

**Bob Frissell**, bestselling author of the 25th anniversary edition of *Nothing In This Book Is True, But It's Exactly How Things Are*

# Introduction

This book revives knowledge that was lost to humanity when Egyptian civilization was destroyed (which occurred more than 5000 years ago), including fundamental knowledge about the structure of our consciousness, its levels, and the laws of transition between these levels. People can use this vital knowledge to activate higher levels of consciousness, helping make their lives more successful and fulfilling. People who have activated higher levels of consciousness are capable of making decisions of a higher level with fewer mistakes. Such people are capable of evaluating the world in entirely different ways, ways that can completely change the course of their lives. You could say, in fact, that they become conscious people. It is difficult to overestimate the importance of this knowledge, which can help people achieve the goal of their existence in this physical world—the development of their consciousness.

Although the book title contains the word "divine," the knowledge it reveals is not a religion, a fantasy, a new age ideology, or a pseudoscience; instead, it is an actual science based on nature's laws. *The Music of Divine Spheres* lays out this knowledge in a simple, interesting, and visually compelling form that any person can easily understand. Things created by nature are so perfect that they can be described and visualized using little to no mathematical formulas. That's why the book contains many pictures and drawings.

Before we talk about consciousness and its structure, however, we need to define concepts. Most people don't set a proper value upon a time in their lives. Therefore, they pay little attention to how they spend their time, to the words they say, and to the meaning they associate with these words. That's why it's easy to deceive people by substituting the meanings of words and concepts. Those who want to take advantage of others often do that for their own purposes. However, people often misinterpret and get confused about different concepts, even without third-party "help," as they either don't know or misunderstand the true meanings of words. That's what usually leads to a mess in their mind, taking them farther away from the truth. The scientific community tends to give strange things even stranger names. Therefore, sometimes people use completely different words, such as "mind," "intellect," "subconsciousness," and so on, to denote consciousness. Nature has no object or process that could be called subconsciousness. For example, there is understanding, there is misunderstanding, and there is a partial understanding, but there is no subunderstanding. Why then should there be subconsciousness? There is no logic in such a name. Mind (usually associated with consciousness) and intellect are entirely different things as well. That's why an intelligent person is not necessarily conscious, and a conscious person does not necessarily have to have an academic degree. You can learn more about this difference from my previous book, *The Civilization of Lost Connection*. For those who have not had a chance to read it, I will briefly explain that intellect is an auxiliary instrument of consciousness. It is like a computer aimed at performing narrowly focused tasks. You can install any programs in your brain, just like on a computer. However, just like in the computer industry, the program might turn out to be malware.

Consciousness allows for reality perception that is undistorted by programs operating in the intellect. Consciousness is an integral attribute of the human spirit and the most important thing a human being has. Our being is defined by consciousness, not vice versa. Consciousness is all that a person can take from this world when dying. You can take neither money, power, nor career, but only the consciousness your spirit has. We perceive the world through our consciousness. It is our consciousness that evaluates what we see, hear, and feel. These senses bypass the intellect and are perceived almost instantly before they can be filtered.

To understand what our consciousness is, we must admit a single postulate: our entire environment consists of matter. In nature, there is no pure energy, information, or consciousness that can be isolated from matter. Our consciousness has its material medium, with its structure. The structure of consciousness influences the way we perceive the world. Therefore, knowing consciousness's structure is crucial because it allows us to answer many of our questions and discover methods that help our consciousness evolve in faster and more harmonious ways. So, the structure of our consciousness is based on the amazing, almost "magical" Laws of Spheres. You will get your chance to learn about it from this book. There is nothing strange about the fact that the structure of consciousness is related to spheres because a sphere is the optimal natural shape. Ancient civilizations knew about it. We can thank their ancient languages for such phrases as "spheres of consciousness" and "spheres of knowledge" because they literally name this shape. It is we who have forgotten it.

The book reveals the secret of a very important symbol that ancient civilizations used to represent human consciousness. To understand this symbol properly means finding the right way to achieve the evolution of our consciousness. It has already become an accepted fact that the average person uses only 3 to 5 percent of their brain, which is a fraction of the full potential given to them by nature. However, little to no one is aware of the fact that everyone has several levels of consciousness. Modern humanity usually uses its lower levels only, without realizing its full potential and thus loses numerous opportunities. Ancient civilizations knew that lingering at lower levels of consciousness could be dangerous; they also knew the techniques and instruments for activating its higher levels. You will find a description of some of them in this book. My experiments with such instruments have proven they are 100 percent effective.

How did I obtain this knowledge? I am a physicist, but I have always been interested in revealing the secrets of the universe and the secrets of our consciousness. Unfortunately, traditional science often fails to answer even the simplest questions. That's why I've spent many years carefully studying sources of information modern scientists usually ignore.

Some of these sources hint at the perfect structure and properties of human consciousness. For example, in his book *The Ancient Secret of the Flower of Life* [1], Drunvalo Melchizedek talks about his conversation with Thoth (Hermes Trismegistus). To get a better understanding of the structure of consciousness, Thoth advised Drunvalo to draw circles, which are projections of spheres, placing them in a certain way on a piece of paper. Having summarized and analyzed the information obtained from various sources, I started exploring spatial structures using geometry, sacred geometry, and modern software tools. The fundamental difference between sacred and conventional geometry is that conventional geometry was invented by people, and sacred geometry was developed by nature itself. So, when you want to understand nature, you must seek answers in its creation. When you dive into sacred geometry, you dive into absolute harmony and flawless perfection. That makes a significant difference. What is created by nature is always optimal, flawless, and perfect. Eventually, I found a secret key. This key is the new, amazing Laws of Spheres, the laws of nature that describe the structure of space, its dimensions (levels), and the principles of transitioning between dimensions. The structure I have found is perfect, simple, and unique—it can be extended indefinitely, and it can be described using an extremely simple mathematical sequence that even schoolchildren would understand. The consciousness of humans, as well as all other living beings, is built in accordance with this structure. It was the structure Thoth hinted at.

The Laws of Spheres gives birth to another important law: the Universal Law of the Harmony of Vibrations, which also influences our perception of the world around us. You will learn how ancient civilizations applied this law, as well as the vibrations of sound and light, to heal people and activate their consciousness. You can use it in just the same way. This is the key that allows one to unravel the mysteries people have been unsuccessfully trying to unravel for many centuries. You will discover the main secrets of the Egyptian pyramids, the nine Egyptian nested crystal spheres, the mysteries of many ancient symbols such as the Sprout of Life, the Flower of Life, the Tree of Life, and many other secrets. In my research, I have found that most things we believe we know about the world are wrong.

For example, remember how we studied geometry at school. We had to memorize the value of pi, the number used to calculate the circumference and the area of a circle. Pi is one of the fundamental values widely used in mathematics and physics. People are confident that this value is absolutely correct, so it has been hardwired into the memory of billions of computers and calculators. Mathematicians have been spending time and resources to calculate two quadrillion signs after the comma in this number. One can say that the commonly adopted pi value has become a symbol of modern science. Pi is depicted on T-shirts and paintings; people carve it on their bodies as tattoos. There are even two international days for pi. Its monuments are being raised worldwide. Poems and movies are being devoted to it; it is being reproduced in musical instruments and used in interior and jewelry design. It has allegedly been found in religious texts. You will be surprised to find out that this value is wrong, just like many other fundamental values. Unfortunately, this is not the only example of human misconceptions about the world. In this book, you will discover the true value of pi and learn about its relationship to sacred geometry and the structure of space.

The notion of the structure of our consciousness originating from the structure of space has made it possible to understand why we feel certain combinations of musical tones, colors, or physical proportions are harmonious, whereas others feel disharmonious. It took me about 2 more years of research and experiments before I discovered that our standard musical scale, or musical octave, is imperfect because it does not fully resonate with our consciousness. I have also discovered the reason why we feel some combinations of musical tones are disharmonious and why they hurt our ears. In this book, you'll find a new inspiring musical scale allowing you to create new music that can help heal people and activate their higher levels of consciousness. However, this knowledge can be applied not only to music but also to light, color, and proportions, thus bringing about new opportunities for people. This knowledge has an immediate practical application. For instance, it might help artists pick up a color's gamma, allowing them to embed certain emotions in their works. This fact alone can revolutionize visual arts. Using specific proportions, you can also create objects and architectural structures, harmonizing and enhancing our consciousness. This knowledge can have many other beneficial uses as well.

This book will also teach you how to make musical instruments that can help you activate the highest levels of your consciousness. These instruments have a positive effect, irrespective of the kind of music you play, because their effect is determined by the musical scale. So, you don't have to be a musician to use them, even though good music will undoubtedly have a more substantial effect. These instruments will give you the opportunity to play and listen to the new music in the new musical scale developed by nature itself. This is the Music of Divine Spheres, the Music of Heaven.

From this book you will also learn how you can accelerate the formation of new levels of your consciousness. This knowledge can be used for the main purpose of accelerating the evolution of our consciousness—therefore changing this world for the better.

# Chapter 1

# Space and Dimensions

*Ancient folios were rescued at the cost of many lives at times. However, it is still amazing how deep and accurately they describe the universe, whereas modern science has got only a little bit closer to understanding some of its aspects. In ancient times there were civilizations and cultures with much broader ideas about the nature of the universe than our contemporaries. A lot of knowledge of ancient civilizations was destroyed by either wars or fanatics of new religions.*
Nikolai Levashov

What is the notion of other dimensions? The issue of other dimensions and worlds has been one of the most mysterious issues to humanity. Even scientists have only a vague idea of what space is, whereas other dimensions are thought to be something inaccessible, something that exists in a completely different or "parallel" world. That's why there are so many various beliefs, interpretations, theories, and fantasies related to this issue. Consequently, it is necessary to settle upon the topic and clarify it a little bit.

To do this, we will have to address geometry. However, modern geometry will be of little help. What we really need is an almost forgotten science exploring the mysteries of proportions and relations between various planes of existence. We need sacred geometry. This mysterious science has been given its name for a reason—because it has an amazing peculiarity. It reveals its secrets to hearts that seek the knowledge but not to intellects. Try practicing sacred geometry, and you'll see it for yourself. In the Russian language, nouns have a gender: geometry is feminine. Like all proud women, like nature herself, she does not reveal her secrets to those who would address her with a closed heart.

We will not only turn to sacred geometry but also rediscover many of its mysteries, learning keys that were lost many millennia ago. We will have to restore some of this knowledge without relying on scientific authorities because there are no authorities in this field.

Before trying to understand what space and other dimensions are, we need to cast off illusions, misconceptions, and misbeliefs we have accumulated over our lifetime, largely due to scientific authorities.

Let's first remember how traditional science explains the dimensions and laws of space. When we talk about space, we need to define a volumetric unit corresponding to it. That is another important thing to consider. Let's start with a point. Let's call it point O. Its dimension equals to zero, so we cannot characterize it by just any number. As all numbers are relative, no single one can be applied without the others. This point simply exists, and there's nothing you can do about it. Digits are powerless here.

We would have a little bit wider variety with space's dimension equal to one. This is a line that contains our point O. This is an infinite line-up of points. We need to apply a variable number to distinguish them somehow.

Therefore, our line can also be applied as a system of coordinates in linear space. If you take the segment line with its length equal to one, then this segment line would become a volumetric unit of linear space. You need to change your attitude toward the concept of "volume." Despite the fact that our intellect is used to consider it as three-dimensional, it can actually have more or fewer dimensions. Its number of dimensions should correspond to the number of dimensions of its space.

You might wonder why there is only a single line. Sure, you can draw the second line through point O, but these two lines together will form an entirely different space with the dimension equal to two — that is flat space. The second line hints at the need for two numbers to distinguish each point in this space. This means that two intersecting lines can be applied as a system of coordinates in flat space. There is an infinite number of lines intersecting at point O at an infinite number of angles in this plane.

Although the scientists chose 90 degrees angle for the coordinate system, we will see later that the most optimal angle between the coordinate axes is 60 degrees. For this reason, scientists took a square for a unit volume of space with a dimension of two, and a cube for a unit volume of space with a dimension of three. Later we will see that the more optimal shapes are those built on the basis of an angle of 60 degrees—an equilateral triangle and a tetrahedron, respectively. A tetrahedron is a pyramid consisting of four equilateral triangles.

Ancient knowledge, which we will talk about a little later, says that volumetric unit should be as symmetrical and optimal as possible; thus, it should be minimal. This last condition in two-dimensional space corresponds to an equilateral triangle, and in three-dimensional space to a tetrahedron. The coordinate axes in this case will be located at an angle of 60 degrees to each other. You know that everything is optimal and necessary in nature. Traditional volumetric shape is convenient for our intellect, but it is far from optimal. As you will see later, applying it as a volumetric unit leads also to incorrect conclusions about space.

Three-dimensional space can be built in the same way. We take the point O as a starting point and draw three lines at an angle of 60 degrees to each other. Then we mark segments with the length equal to one on each of them. The lines containing these segments form three planes extending infinitely and intersecting at point O.

You are probably wondering what is next. How do we describe a four-dimensional world? We can do that by just extending the analogy within the mathematical line of reasoning. After all, we can say that a single-dimensional world is an infinite structured set (or multitude) of zero-dimensional worlds embedded in it. The two-dimensional world is an infinite structured set of embedded single-dimensional worlds. The three-dimensional world is an infinite structured set of embedded two-dimensional worlds, and so on.

As you can see, spaces have another curious peculiarity. Each space of a smaller dimension can be entirely embedded in the space of a greater dimension. Additionally, their structure is arranged in such a manner that all spaces contact each other at the initial point. This means that this point belongs to all spaces at the same time. Since we have chosen a random point, we can argue that it is possible to get into any space from any point. In other words, every point we take belongs to all spaces at the same time. This means that to transition to another dimension, you need no special places. You can do it from the very place you are at the moment.

Unfortunately, science has not gone further than such a simple, at the high school level, explanation of spaces. There are, of course, different theories, but none of them has anything to do with reality. The reason for this is that several centuries ago, science, like religion, was separated from spirituality, which had given it a connection with nature and the real world. Therefore, modern science, although it can conduct experiments, cannot explain even elementary things. Scientists will not be able to answer, for example, whether there are spaces with dimensions of four, five, or more; is it possible to travel from one space to another; how our body will change if we move to a space of another dimension. Science can only say that in order to represent an object in spaces of different dimensions, it is necessary to construct projections of this object in the corresponding spaces.

As we remember from geometry, to switch from a three-dimensional shape to a plane, we need to project this shape onto a plane. This is similar to the way a three-dimensional shape is usually drawn on a sheet of paper, or an object's shadow is projected on the ground. That is, we will transition from a three-dimensional space to a two-dimensional one after we draw perpendicular lines (as they are the shortest and optimal segments) from the points of a three-dimensional object to the plane, and then draw an outline of the resulting shape in the plane. We see that the first required condition is a 90-degree angle. Why do we need an angle of 90 degrees? Because this angle allows for the optimal and the smallest projection. Similarly, we can switch from a two-dimensional space to a single-dimensional one by projecting a flat shape onto one of the axes of a two-dimensional coordinate system — because an axis itself is a single-dimensional space.

In general, we can say that N-dimensional space is a section (or projection) of space of dimensionality N+1. Our brain reads images from our eye retina in a similar way because these images are also a section, a projection of three-dimensional space on the surface of our retina. When space loses its dimensionality, it becomes a projection containing less information. That's why you don't know what's behind an object when looking at it. You don't have that information, even though it's available in a three-dimensional world.

According to modern science, if you manage to move from three-dimensional space to lower-dimensional spaces, then your body will turn first into a flat object, then into a straight line segment, and then into zero. Zero means nothing. That is, you will completely disappear. It's like a one-way ticket. If you are on a plane, you won't be able to come back to the 3D world. If you are on a line, you won't be able to come back to the plane. If you disappeared, then you disappeared forever. As you can see, applying school knowledge allows you to calculate everything in a simple, fast, and...wrong way. Our calculations will be correct from the point of view of regular geometry; however, they will be wrong from the point of view of real spaces. If the way scientists imagine the world turned out to be true, then Reality would have disappeared long ago. Since, despite all the efforts of scientists (just kidding), reality still exists, then apparently it obeys other laws.

Ancient knowledge teaches that during transitions from space of one dimension to space of another dimension, certain conditions and rules of the real world must be followed.

First condition: we cannot jump through emptiness. We must leave a continuous trajectory, or an imprint, in all intermediate spaces while transitioning between them. Therefore, each subsequent projection should have at least one point of contact with the previous one. This is the principle of the spatial continuum or connectedness.

The second condition may not seem so obvious, though it has a simple explanation. The dimensionality of space can be decreased or increased by applying two attributes necessary for such transformation: a 90-degree angle and a 1 unit-length segment. If you remember the Pythagorean theorem, then you will understand why we need these two constants. If you follow these rules, you will be able to travel back and forth between dimensions. Besides, there is no zero in nature.

So, traditional science not only cannot answer elementary questions, but also leads you to a wrong understanding of the world. In contrast, ancient civilizations and their ideas about the world were more in line with reality. Their connection with nature was much stronger than ours, so their knowledge describes the real world better than ours does. The problem is that almost no material evidences of this ancient knowledge have been preserved. No wonder people forgot about them. In this sense, Drunvalo's books are unique—they describe the author's personal conversations with a living bearer of the knowledge of the ancient mysteries of Egypt, namely, with Thoth, also known as Hermes Trismegistus. At different times, some people considered Thoth a god, others just a sage. One way or another, he possessed, from the point of view of ordinary people, supernatural abilities. As it turned out later, Thoth was also the one who built the Great Pyramid. Carefully analyzing the content of his conversations with Drunvalo, some of which are described in the book *The Ancient Secret of the Flower of Life* [1], one can immediately notice that Thoth is indeed the bearer of ancient knowledge that humanity had lost. Every word uttered by Thoth made sense and was carefully weighted. He didn't say anything unimportant. His every single word was important and matching the place and time it was uttered. We should learn this from him, as well as many other truths.

In conversations, he gradually pushed Drunvalo to get understanding of the important law that determines the structure of space, as well as the consciousness of living beings. However, Thoth never presented Drunvalo with ready-made solutions, telling him instead that he should figure it out for himself. He gave only hints, trying to guide the interlocutor's thoughts in the right direction. This makes a lot of sense. Knowledge will remain in your mind only when you fully understand it through your own efforts. Therefore, the main truths remained undisclosed. We can see the manifestation of an important law in this: nothing should be given to humanity, cut and dried; otherwise, it will not remain accumulated in their consciousness. If they want to obtain it, humankind itself must obtain and experience the knowledge.

Fragments of ancient knowledge, given by Drunvalo from the words of Thoth, formed the basis of my research on this subject. It was these clues from Thoth that allowed me to find the key and fully restore the lost ancient knowledge that Thoth hinted at. Moreover, this key also allowed me to find other laws derived from it, which you will learn about a little later in this book.

But first, let's take a brief look at the information about space and the structure of consciousness we have received from Drunvalo and Thoth. This information will serve as a starting point for our research. Let's start with the ideas of the ancient Egyptians about the creation of the world.

Imagine that you're in the beginning of the creation of Reality. You're the spirit flying in the middle of the Great Void. There are no things, no forms, no time. You even don't have a body. You are just spirit, pure consciousness. You would get bored, and would want to try something new. So spirit shoots a beam of consciousness out into the Void and places some field or energy at the end of this beam separating a point in the Great Void. Sacred geometry started here. Now the spirit will use the distance to that point as a unit of distance in the space. Then, the spirit rotates the beam around it in all possible directions that creates the first sphere. The spirit is now sitting in the center of the created sphere in the middle of its first creation. Eventually it'll make a decision to do something else. Spirit decides to make its first motion. So consciousness goes to the surface of the sphere and projects another sphere exactly like the first. This is the first day of creation. When it does that, it has done something very unique and special in terms of sacred geometry. Two intersected spheres have created two new forms—a flat membrane surrounded by a circle and a three-dimensional vesica piscis. Then, on the second day of creation, spirit lands on the innermost circle point and projects another sphere. Then spirit starts moving around the equator of the original sphere and projects new spheres until, in the sixth day of creation, it gets the perfect pattern—the first sphere precisely surrounded by six other spheres. Spirit continues the process of creation by moving to the innermost points of created new shapes and projecting other spheres, producing more and more complex shapes. Drunvalo says that this sacred geometry is not just mathematics, and it's not just circles or geometries. This is the living map of the creation of all Reality. You must understand this or you'll get lost and won't understand what this book is about.

According to Thoth, different living beings have different levels of consciousness, and the proportions of their bodies correspond to their levels of consciousness. That is, by measuring the proportions of the body of any living being, you can determine its level of consciousness. Human consciousness, for example, is between the fifth and sixth levels. The consciousness of living beings, as well as the consciousness of entire planets, can be measured also by measuring the size of their active outer energy shell. In addition, the consciousness of our planet is directly related to the angle of inclination of its axis. He also said that one is actually not one, but the square root of one; and the squaring of the circle is related to consciousness and the golden ratio.

According to Drunvalo, each human being has an energy structure in the form of a star tetrahedron, and this structure is differently oriented in a man and a woman.

In order for Drunvalo to understand the structure of consciousness, Thoth advised him to draw patterns from circles of the same size on a piece of paper. By doing this, Drunvalo has got many interesting patterns and geometric shapes, including the Egg of Life, the Flower of Life, and the Tree of Life. Then, examining the geometric shapes, Drunvalo found confirmation of them in nature, namely in the spatial shapes formed during cell division. Interestingly, by doing this, Drunvalo was only one step away from discovering one of the greatest laws of nature, perhaps the most important law in the history of mankind—the main Law of Spheres.

## Chapter 2

# Spheres and Consciousness

*The fundamentals that lie closer to the source are known by God only and by those people who are friends to God.*
Plato

You might find it difficult to believe that our consciousness is composed of interconnected spheres. However, ancient civilizations knew about this fact. The only difference for different people is in the number of spheres and their "accumulated contents." The expression "spheres of consciousness" did not appear from nothing. It has a solid grounding in fact. Now, you have the chance to make certain of that. This chapter will reveal the main Laws of Spheres.

The consciousness of living beings evolves by accumulating new spheres and connecting them to those already accumulated. In this way, living beings accumulate their evolutionary potential. This process is natural because information and experience are always accumulated within the material medium. A sphere is the optimal natural shape for such a medium. The more spheres are collected, the more levels there are within consciousness or the higher its dimensionality. In other words, dimensionality depends on the number of accumulated spheres. However, this process is not as linear and simple as, say, adding balls to a basket to increase its mass by a certain amount at a time. No, it's completely different. In this case, we are dealing with much more curious laws related to energies, levels, and active external shells rather than to their volumetric and weight characteristics. A certain number of spheres accumulated activates the energy surface of a certain outer shell shaped as a sphere of a greater diameter. An outer shell is activated. If you could see it, you would see a glowing ball exactly fitting to a cluster of identical spheres of a smaller size as one may see on the domes of Tibetan temples.

Let see how one can make transitions between different dimensions in the space of spheres and space in general. Imagine ten spheres of diameter equal to one unit, stacked together in a large tetrahedron (which is a pyramid with four vertices)—six spheres at the bottom, three in the middle, and one at the top, Fig. 1. For simplicity, only three of ten spheres are shown in the drawing. Pay attention to the fact that the lateral edges of the large tetrahedron make a tetrahedral coordinate system with the center at the point O. We have constructed one of the shapes of the space of spheres. In our example, we will see what happens when we transition between spaces with the AB segment. However, the simple mathematical process we are going to apply can be attributed to folding or expanding whole multidimensional shapes, that is, to their transition to lower or higher dimensions. For example, you will see how a tetrahedron will "be folded" into a triangle, then to a line, and finally to a point.

Let's, knowing the ancient rules, build the right trajectory of our movement in spaces, that is, find the projection of the edge AB to the point O, successively building projections and moving from spaces of higher dimension to spaces of lower dimension, without interrupting the trajectory.

We will start by obtaining the BD segment that touches the original edge AB at point B and forms an angle $\alpha$ of **30** degrees with it. To frame this projection, we've applied the right angle and 1 unit-length segment for the first time. At the second step, we obtain the next DE segment, forming an angle $\beta$ = **35.264** degrees with the previous BD segment (from this point onward, for simplicity, I will round up endless fractions to three decimals). As you can see, **the angles of transitions grow when we move to lower dimensions** and approach point O. You must note this important regularity.

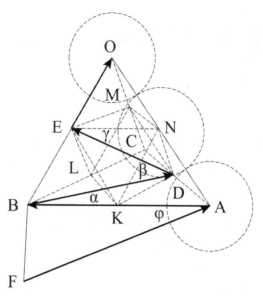

Fig. 1. Space of spheres and transitions between dimensions using the tetrahedral system of coordinates.

Next, we get the projection EM, which forms an angle γ of **45** degrees with the previous projection DE. Finally, the desired projection of the edge AB to the vertex O will be a 1 unit-length segment MO, directed to the previous projection EM at an angle of **60** degrees. It should be noted that in this last step of constructing the sequence of projections, we could make our trajectory shorter by one step. Without violating the rules, we could immediately build a projection from point E to point O, that is, a projection of EO, at an angle of **90** degrees (angle DEO). Therefore, near the point O, the trajectory of our movement looks a little strange. It can make a turn through one or two angles of 45 degrees at once (that is, at an angle of 90 degrees) and merge with 1 unit-length segment, with the other end falling directly into our point O. In fact, there is nothing strange in this, since a small tetrahedron OEMN, as you will see later, behaves like a single object. This "strangeness" will disappear if we choose as the starting point not the point O, but a point lying in the center of an edge of the tetrahedron, for example, the point M or the point N[1].

Therefore, the apex of tetrahedron O isn't the best starting point of zero-dimensional space. It is more reasonable to choose the middle of its edge; point M, for instance. This will allow us to get rid of uncertainty without affecting our trajectory. You will see later that points M and N belong to two-dimensional and single-dimensional spaces, and point O belongs to three-dimensional space. Therefore, this explains the strangeness of our projection near this point. But for now, let's just take it for granted.

Let's arrange all projections according to their size and the order in which we have obtained them. We must also consider the fact that two actually equals the square root of four, and one equals the square root of one. As a result, we get a mathematically precise numerical sequence:

$$\sqrt{4}; \sqrt{3}; \sqrt{2}; \sqrt{1}$$

As we can see, this sequence has no number that is equal to zero, which is understandable. There is no absolute zero in nature, as there is no absolute emptiness. Nothing can be created of an absolute zero, and nothing can pass through an absolute emptiness. Therefore, the projection of our singular three-dimensional segment on a point (in the zero-dimensional world) is not a zero, as we were taught in school, but one. There's another interesting detail. Projection sizes directly point to the dimension of space they have been derived from because they have been obtained by square rooting the dimension of that space—that is, the size of an object points to the dimension this object belongs to. Isn't it amazing how the world works? Each object in different dimensions has a different size! It's easy to imagine. When you dive under the water and open your eyes, all objects and distances seem different. However, objects and distances in reality remain the same. Such an effect is caused by the different density of the environment a person has got into. So it is not about the objects but about the way we assess them in various conditions. Space of another dimension can be figuratively compared with such an environment.

By extending the sequence to higher dimensions, we can see that our original singular AB edge is only a projection of a certain larger segment from the fourth dimension, which equals to the square root of five. In its turn, this segment is a projection of an even larger segment from the fifth dimension, equal to the square root of six, and so on. As a result, we get an amazing series of projections:

$$\sqrt{1}; \sqrt{2}; \sqrt{3}; \sqrt{4}; \sqrt{5}; \sqrt{6} \ldots$$

Using these rules, we can even define the angle between this segment from the fourth dimension (FA segment) and the three-dimensional AB segment as they intersect at point A. This angle equals to **26.574** degrees. Therefore, the angle of 26.574 degrees is the angle of transition between the third and the fourth dimension.

That's not all we can say about the fourth dimension. The size of the segment in the fourth dimension is determined by a well-known proportion of $\Phi$ (read as "Phi") equal to 1.618...This is the base for almost everything that nature created in the three-dimensional world, including your greatly respected body.

You can see we have been given this proportion from the fourth dimension. That is someone who lives there has your three-dimensional body that lives in our third dimension. This person is you. That is why we perceive the objects that have been created within this proportion as beautiful and harmonious. Despite the objects belonging to our three-dimensional world, the feelings they cause belong to the fourth dimension. However, this strange proportion has a different value in the dimension it originates from.

The formula for calculating $\Phi$ can be written as the following:

$$\Phi = \frac{\sqrt{5}}{\sqrt{4}} + \frac{\sqrt{1}}{\sqrt{4}}$$

This formula will allow us to get a better idea of where we stand at the moment while watching over $\Phi$. This place is determined by the number in the denominator, namely square root of four, which means we are in the third dimension. You can calculate the value of $\Phi$ for each dimension with a similar simple formula. That value will be different in each of them.

These are far from being all the surprises the golden ratio conceals. Ancient Greeks called it "divine" for a reason. It has extraordinary properties reflected in various formulas (e.g., $\Phi^2 = \Phi+1$ and other unexpected formulas). We can say that this ratio is a mathematical symbol of the fourth dimension. Its geometric symbol is a five-pointed star or a pentagram. It is a proportion of $\Phi$ in its purest form! This proportion reveals itself not only in the ratios between the segments inside the star. If you draw a regular pentagon with an edge equal to one, the distance between its opposite vertices will be equal to $\Phi$, Fig. 2 (left).

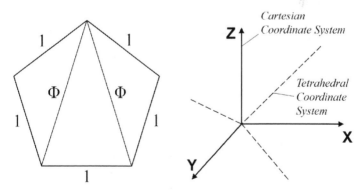

Fig. 2. A pentagram as a symbol of the fourth dimension and «Phi» (left) and the relation between an optimal (tetrahedral) and a rectangular (Cartesian) coordinate system (right).

The five-pointed star is formed by five segments equal to Φ. This last condition gives us another exact formula for calculating Φ and two other special angles related to it:

$$\Phi = 2sin\left(36^{o}\right) = 2cos\left(54^{o}\right)$$

We can exit the single-point space at the same angles as we have entered it. We have entered single-point space O at a right angle. Let's see where right angles can take us after we leave the point behind. If you draw a perpendicular line between point O and each of the tetrahedron's lateral edges, you will get...a rectangular, or as it is also called, a Cartesian coordinate system. But it's not just a well-known traditional system. The relationship between these systems of coordinates is much more profound. The tetrahedral coordinate system axes are the bisectors of the rectangular coordinate system, Fig. 2 (right). Thus, all the time while we were framing our trajectories within the coordinate system with such strange angles between its axes, we were within a rectangular coordinate system at the same time, without even realizing it. Everything is surprisingly interconnected in nature. Has this already made you feel excited? Just wait. It's going to be even more exciting!

Can we practically witness the reflection of the pattern we have just obtained? Take a sheet of paper and draw all the segments (joining at one point) on it according to their size, the angles between them, and within the order you have obtained them. If you then connect their other ends with a smooth line, you will obtain a spiral rising from infinity and reaching toward the point. Now draw a perpendicular line through the sheet of paper you have drawn your segments on as it "pierces" the paper at this very point. The line and the spiral together frame an approximate trajectory of the matter movement as it passes through the folded space of the black hole (if it acquires enough energy to get out of single-point space, having almost turned into light). Therefore, when matter passes through the black hole, its dimensions get folded and then expand again.

Since space exists physically in various dimensions, space of our consciousness can also exist in various dimensions. So, we have humbly approached space of mind or space of consciousness. Like any other space, it can have different levels or dimensions. Consequently, there are ways to transition between them.

You might be surprised. "What do geometry, projections, and angles have to do with it?" you might ask. "What does all this have to do with consciousness?" Everything mentioned above has a direct relationship with consciousness. The fact is that spaces, dimensions, and transitions between them are subject to the same laws regulating the structure of our consciousness and the consciousness of other living beings. These laws are called the Laws of Spheres. They manifest when equal-diameter spheres connect with each other to create clusters of specific spatial shapes. The Laws of Spheres are related to the tetrahedral system of coordinates and the projections and transitions we discovered above. You will soon see it for yourself.

Let us now draw our attention to what we have in terms of spaces and our ability to perceive them. If you ask different people about the dimension in which they live and the dimension they go after death, many of them will probably answer they live in a three-dimensional world (we commonly associate it with a three-dimensional rectangular system of coordinates), and the afterlife will transition to a fourth or even higher dimension. No one can provide a precise description, however, because no one actually knows what it looks like and where it is situated. If you further ask them to explain how they understand two-dimensional and single-dimensional worlds, they will likely answer that the two-dimensional world is a plane and draw two perpendicular axes, X and Y. At the same time, they will see the single-dimensional world as a line. Logically, the zero-dimensional world is a point. If you ask people whether they see lines and points, they will probably look suspiciously at you (have you lost your mind?). Then they will answer that, naturally, they can see them because they are not blind. In fact, most of these ideas and concepts are nothing but illusions. Many people around you have had a hand in their development, as well as books, teachers, and even your own intellect.

Let's start with the fact that you can see neither lines nor points, because both lines and points are entire worlds of smaller dimensions (they are less informational). You draw lines and points with pieces of planes of a certain length and width, and your schoolteachers have convinced you that you draw and see lines and points. But you can't see the three-dimensional world either, so you don't know who's hiding behind the nearest bush. All you can see are pieces of the plane, and this feature of yours is subject to the properties of your eye, namely the superficial structure of its retina. You can see only a limited projection of the three-dimensional world on a plane, and your brain is like a computer. It calculates the dynamic patterns of a flat layout, allowing you to see the illusion of three-dimensionality. But this is a pseudo-three-dimensional vision, not a real one.

Your hearing is linear. It measures air pressure at two points, and your brain "calculates" the dynamic pattern of the pressure, just like in the previous case. Then it obtains additional information out of it and creates the illusion of three-dimensional sound. So, if you are blindfolded, you can't tell whether it is a real person talking next to you or the sound is coming from two speakers at entirely different places. The sense of smell is a single point. This limitation of our perceptive apparatus generates various misconceptions and illusions.

What do we really have? Not much. There's no reason to turn up our noses. We have a set of single-point, segmental, and piece-and-plane senses. Our brain uses these snippets to compose a certain picture of the world around us. The only three-dimensional thing we have is our bodies, which are an important link, a bridge connecting us with the fourth and higher dimensions. Our other bodies are located there. We also get our higher-level senses from these dimensions.

This is an extremely important point humanity does not understand and therefore attaches no proper importance to. Instead of using their bodies and minds as exceptional instruments of activating higher levels of consciousness to step up to higher worlds, most people, in contrast, use their intellect instead of consciousness and lose at least two real dimensions from their lives. Sometimes, they even destroy what they have accumulated in the course of their evolution in past lives. This is the result of focusing on intellectual thinking. When a person does so, the feelings coming from the third and fourth dimensions fall out of one's life completely. Most people think continuously.

One day, you'll leave this three-dimensional world and abandon your physical body and the "mathematical" apparatus of your intellect. Many physical senses will disappear from your life. You might have to face a two- or even single-dimensional world. Considering how easily people's minds are captured by empty forms and scenes, you might get stuck there forever unless there's someone from higher dimensions to rescue you. Someone who has promised to help you? If you've read Dante, you can recall what was written above the gates of hell, as well as another interesting detail. In his *Divine Comedy*, Dante described hell as a downward spiral with a swirling storm inside. It was taking away the souls of the people trapped there. There were several layers (or circles) in this spiral. Each of the circles was devoted to a certain group of souls (by match).

If you think about the spiral and the levels of space we have just discussed, you will understand what he was talking about. Therefore, if you fail to learn how to enter higher dimensions in the course of your life, you need to at least learn how to avoid degenerating to lower ones. First of all, it is necessary to learn to prevent your intellect from being easily caught. To do that, you need to acquire attentiveness and elevate your consciousness.

An important thing needs to be said here. Many people believe consciousness can be evolved through meditation and detachment from the world. This is a misconception. Meditation is only needed to gain control over the intellect, to stop the process of continuous thinking. According to Nicolai Levashov, human consciousness evolves only in the process of practical creativity aimed at changing the world for the better, bringing it in line with natural laws of evolution, and fighting all the faces of evil.

Ancient Egyptians described such creativity as "serving the gods." Now you know what this notion actually means. Unfortunately, in the course of history, it was substituted with religious rituals and meditations. In the process of creativity, you gain the experience and practical knowledge of natural laws. They accumulate in your consciousness while increasing its dimensionality, raising and "pushing" it to higher levels.

According to Nikolai Levashov, the results of actions aimed at so-called self-improvement are negligible in terms of their scale and load. New qualities and properties in a person cannot appear without an appropriate load. Evolutionary acquisitions arise only when the actions of the person following the spiritual path are aimed at the benefit of others. Actions aimed at the benefit of others, in terms of their tasks and loads, can reach the scale of the Universe. In the process of solving such problems, flows of matter through the actor may reach such a level that they change the actor himself. Of course, the level of tasks to be solved must correspond to the capabilities of the actor.

If a person does not live following natural laws, if one lives against them, then consciousness does not evolve. It degenerates. This process can be compared to the process of accumulating errors in the system. If there is a critical number of errors, the system fails. When degenerating, the human spirit can also cause a genetic degeneration. This, in turn, can push a human being to lower evolutionary levels. You can find many examples of this yourself. Egyptians knew about the necessary conditions for the right evolution of people's consciousness. However, this knowledge was artificially distorted to make people passive, ignorant, tolerant of evil, and easily controllable.

Now you will see how "magical" the Laws of Spheres are, how they are related to the tetrahedral coordinate system, as well as to the projections and transitions that we obtained above. We will try to compose a space of tangential spheres of equal size to better understand why the accumulation of spheres triggers the activation of new levels of consciousness and why this does not happen according to a linear law. Let's assume that the very first, or initial, sphere we take to compose space is fixed at its central point. When we connect other identical spheres to it while trying to form a cluster, and when this cluster spins around the initial sphere's center, we see that rotation makes centers of the added spheres slide in space along spherical surfaces, with diameters exceeding the diameter of the initial sphere. All these larger spheres (produced by a rotation of the cluster of identical spheres) will be nested or embedded in each other, meaning they will have a common center.

For example, you may add three more identical spheres to the initial sphere to compose a tetrahedron of four spheres. Then you may rotate this figure around its initial sphere's center to make centers of the added spheres slide along the same spherical surface with a twice larger diameter than that of the initial sphere. This new surface, with its radius equal to one, denotes the first layer of spheres — and at the same time denotes the first level of consciousness and zero dimension of space of consciousness. "Zero" here means just a conventional number or a label, which does not mean that this dimension is equal to zero. We call it "zero" because this shape (a small tetrahedron) has a single possible distance between its vertices, which is equal to the initial sphere's diameter. There are no other distances and levels. That's why this figure seems so mathematically strange, just like the projections we have been framing before. We can mathematically consider the initial sphere to be a zero level of a zero dimension, but practically this dimension exists beyond all other dimensions. Same as spirit.

If we will continue adding spheres, the cluster will grow; and new levels will appear. Here, we need to define the notions of layers and levels. The point is that not all larger spheres encompassing the cluster shape have a diameter divisible by that of the initial sphere. The framing sphere, with a diameter divisible by that of the initial sphere, is called the layer sphere, and the spatial zone between the nearest of such spheres is called a layer. Other nested spheres are called levels because they do not comply with this rule, Fig. 3. Each level belongs to its appropriate layer.

One of the unique Laws of Spheres holds that all layers, starting from the second one, contain intermediate levels, with their quantity varying from layer to layer. This quantity increases by two at a time when the layer number increases by one. The next layer cannot be composed until the previous one (including all its levels) is filled with small spheres. The following example helps clarify the fact that only a few sphere diameters are divisible by that of the initial sphere. If you compose a pyramid of five small spheres with a square at its base and then rotate this shape around one of the square's vertices, you will get both the first and the second levels. The first level corresponds to the first layer. The center of the sphere located on the square's diagonal will move in the second level. At the same time, the external active shell of the new shape will jump to the next level, with the largest radius possible for this shape. This radius is equal to the square's diagonal. However, it is not divisible by the initial sphere's radius because the square's diagonal is not divisible by its edge (one). It equals the square root of two (see Fig. 3).

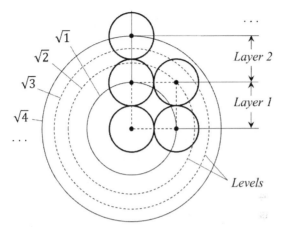

Fig. 3. Layers differ from levels in that the diameters of the layer spheres are multiples of the diameter of the original sphere.

Because levels don't change smoothly but discretely, practical transitions between levels of consciousness are not smooth either, but rather discrete. Because human consciousness is occasionally capable of such leaps, when these transitions occur, they are usually thought of as "enlightenment." When a person has developed one's spheres of consciousness to a certain level, a small change or external stimulus is often enough to allow one's consciousness to ascend to the next level. This transition is an important process for a person and one's entire life. Ancient Egyptians realized it well, so they developed special methods to stimulate such shifts. This was one of humanity's greatest discoveries. Unfortunately, it was lost.

The word "layer" may make you think the initial sphere gets covered by the first layer of spheres all around it. Then the second layer "sticks" to the first layer evenly, and so on. But it's not that simple. It's much more interesting.

The fact is that spheres compose clusters of optimal shapes — that is, shapes with the distances between spheres as short as possible. Imagine that each new sphere is influenced by the force of attraction to a cluster. It "slithers" into the deepest place on the surface of the cluster to take the closest place to the common center. It is these "closest places to center" that we'll call the innermost cluster points. So, the new sphere moves to the innermost cluster point. If this condition is met, the cluster shape will be as optimal and sustainable as possible. Therefore, when the distances between the centers of all spheres are the shortest possible, such space can be called the optimal space of spheres. This is quite natural, for all things in nature are optimal and strive for the most sustainable and strong connections.

Practical experimenting with spheres will undoubtedly help you obtain an easier and better understanding of this process and its evolution. The best thing you can do is get a set of identical balls and try to stick them together into spatial symmetrical, optimal shapes. This should certainly help you, especially if you have difficulties imagining all this in space. For example, when you stick more and more balls together, you will get primitive optimal shapes at first. Then, as new balls are added, the cluster will become larger, while its shape will be closer and closer to the ball. Soon enough, you'll see each new layer of spheres, giving you an entire set of new dimensions. I call it a set because, starting from the second one, each layer has several possible distances between the centers of spheres composing the shape (see Fig. 3). However, the maximum distance between the spheres within the cluster does not exceed this layer's largest size. The higher the layer's number is, the more intermediate levels it contains.

Now, we will look a little bit ahead and see that the dimensionality of each cluster of optimal shapes is defined by (1) the number of possible distances between the centers of spheres composing this cluster and (2) the number of possible seamless transitions between its levels. We will use this as a criterion for defining the dimensionality of shapes composed by spheres.

Thus, when applied to spheres, dimensionality 1 calls for two levels and two possible distances between centers of small spheres within the shape (such as in a square or an octahedron). Dimensionality 2 calls for three possible distance values within the shape (e.g., a large tetrahedron).

Therefore, three spheres placed within an equilateral triangle do not compose a complete or the largest possible shape of the first level in the zero dimension. We can complete these three spheres with the fourth one without changing the dimensionality. The distances between the centers of all spheres will be equal to one — that is, the diameter of a sphere. Thus, the largest possible shape of the zero-dimensional space is a small tetrahedron. This is the shape of the first layer of spheres.

The largest possible shape of a single-dimensional space is an octahedron consisting of six spheres (e.g., the KLEMND octahedron in Fig. 1, composed of central points of edges of a large tetrahedron). There are two possible distances between the centers of the spheres forming it. Therefore, this shape belongs to the first level of the second layer of spheres. This first level of the second layer is actually the second one if we count from the space origin. Or it is the third one if we consider the initial sphere to be the first level. Two-dimensional space corresponds to the large tetrahedron OABC (10 spheres), although it is not the largest possible complete shape of the second layer, as you will see soon.

Fig. 4 illustrates the evolution of flat and volumetric shapes composed of spheres. This evolution of shapes escaped the attention of people engaged in sacred geometry across various centuries. Therefore, they failed to find the main relationships. Fig. 4 clearly demonstrates that space originates not from a separate point or, let's say, a sphere in its center, but rather from the entire primitive shape composed of spheres. This shape accumulates spheres around it while adhering to volumetric symmetry. Therefore, this figure also helps us understand which spheres or points of the large tetrahedron can be considered as the starting point of space. We can see that it is natural to start shaping space from the center of the edge of a large tetrahedron. This point simultaneously belongs to zero dimension (one sphere, equilateral triangle, and a tetrahedron) and the first dimension (a square and an octahedron). The vertices of the large tetrahedron belong to the next, second dimension.

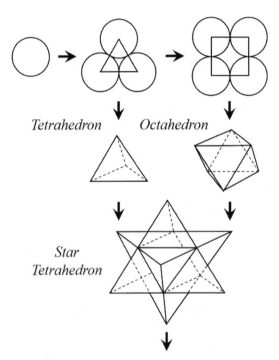

Fig. 4. The evolution (the increase of informational contents) of spatial shapes underlying the spaces based on one sphere, triangle, and square.

Spheres and Fig. 1 help us realize that the tetrahedral coordinate system is developed by nature itself, so it naturally originates from the Laws of Spheres and, therefore, is optimal.

Please note another important pattern of our method because it will become useful soon. Just view the projections we have obtained from a slightly different angle. We have started from the AB segment because it is the largest segment of the large tetrahedron. Then we moved to the BD segment—its shortest segment. In turn, this segment connects to the DE segment—the largest segment of the shape in the previous space of a lower dimension. The octahedron belongs to that space. Then, we moved from the diagonal of the KEMD square to its EM edge, which is its smallest size. While moving through spaces, we were passing through layers with the largest and smallest sizes. Moreover, the largest size of the spatial shape of dimension N is connected to the smallest possible size of the next spatial shape of the dimension N + 1, where N is an integer number (whole number) that can vary from one to infinity. This is an important law on the structure of spaces and their layers. It provides a connectedness of spaces and a seamless transition from one dimension to another.

Thus, all these shapes (a triangle, a tetrahedron, a square, an octahedron, etc.) are composed according to the rule of sphere accumulation. They are accumulated in layers, with each of them having exact upper and lower limits.

In general, there are three methods of building an optimal sphere space. It depends on the most primitive shapes you will choose as its center or basis (see Fig. 4). All possible methods have some common properties. Later on, we'll explore all these methods and investigate their results. We must also remember that we are dealing with animate and conscious objects, not with regular geometry. Geometry only gives you a better understanding of the laws of nature.

Now, let's look at the evolution of flat shapes, as depicted in Fig. 4. We will also study the spaces built upon each of these shapes.

The evolution of shapes takes place within the following sequence. One sphere goes first, followed by two and three spheres (a triangle), and a tetrahedron composed of four spheres. These shapes correspond to the zero dimension. You will see a different picture if you add two more spheres. If you leave them in space, six spheres will become tangential to each other because they are drawn together by forces of mutual attraction. They will compose an octahedron—the optimal shape closest to the shape of a sphere and consisting of three squares or a square and eight equilateral triangles (depending on the points you will choose to highlight). Thus, we have an object called the Seed of Life, Fig. 5 (upper left). Within the concept of the space of spheres, this object is a shape of the first dimension. There are only two possible distances in it: the diameter of the sphere and the diagonal of the square composed by them.

The Seed of Life was composed as an object of the first dimension that "gave birth" to sapient life. This object (an octahedron) reveals a new dimension belonging to the first level of the second layer—without filling it completely, though. The second layer can be completely filled if its largest diameter is equal to two diameters of the original sphere. The octahedron "fills" the second layer only partially because the diagonal of its square reaches only the first of three possible levels of the second layer.

To avoid the confusion and misconceptions that we've been taught in school, we need to understand that actual dimensions are different from those our intellect is accustomed to. The problem is that our intellect is used to perceiving and interpreting visual shapes as flat or three-dimensional ones. All shapes composed of spheres look three-dimensional—that is, our intellect interprets them as such. However, we are not dealing with ordinary mechanical shapes here. These shapes compose the structure of consciousness of animate objects. Each of them consists of its own set of dimensions or possible distances between spheres composing it.

Fig. 5. The evolution of the simplest spatial shapes: the transformation of a tetrahedron (upper left) into the Seed of Life, which is an octahedron (upper right), and the Sprout of Life, which is the star tetrahedron (bottom).

From Fig. 1, you can see that if we take, for example, a sphere centered at point M as the initial sphere (the center of the space), then this sphere will belong simultaneously to the zero dimension (to itself and a triangle) and to the first dimension (a square).

But that's not all. The spheres in Fig. 1 also belong to a large tetrahedron, which is the shape of the next, second dimension! Therefore, we can observe the implementation of the law of inclusion (when a shape of lower dimensionality is fully included into a shape of higher dimensionality) and the law of spatial connectedness.

We must complete the second layer to get the largest completed shape of this layer, revealing the second and the third dimensions. We must add four spheres to the octahedron to create a large OABC tetrahedron, which will be exactly twice the size of the original, small one, and then add four other spheres to adhere to the volumetric symmetry. In this way, we will complete a star tetrahedron (Fig. 4), consisting of two large tetrahedrons. These tetrahedrons should be turned in opposite directions and embedded in each other. The star tetrahedron consists of 14 identical spheres. Memorize that number. It will give you a better understanding of Egyptian mythology. According to their myths, Seth split Osiris's body into fourteen pieces after murdering him and then scattered them around the world. In fact, he split his consciousness, not his physical body.

That was how the Seed of Life "sprouted," giving birth to the Sprout of Life that has grown from the first into the third dimension. It is nothing but the star tetrahedron. Note that the Seed of Life and the Sprout of Life images you can find in existing informational sources are incorrect. You already know the reasons for that.

There are probably no rules that don't have exceptions. The thing is that the star tetrahedron is a specific shape. It contains all dimensions, from the first to the third one and even the fifth one, but there is no intermediate fourth dimension. This very fact makes it an imperfect shape. Therefore, it breaks the perfect level connectedness within the optimal shape, which makes it an exception to the common rule. This has negative consequences for human beings, but you will learn them later.

The third completed layer of spheres will result in the shape of the fourth to ninth dimensions, as you will see below. Spheres allow us to visualize and practically compose it. Its vertices will be in the centers of spheres forming the third layer.

Let's summarize the information we've got. Note that each optimal shape can be accurately embedded in the outer sphere of a corresponding diameter. The framing sphere is the largest one in this space of consciousness, or at this level of consciousness. The reverse is also true: each active outer sphere of consciousness corresponds to a precise geometric shape embedded in it. The more levels, or layers, there are, the more vertices the embedded shape has, the more spherical is the shape of the outer layer, and the greater is the radius of the outer framing sphere.

Each new level allows the framing sphere of consciousness (active outer shell) to expand, increasing the dimensionality of space and its information content. However, it does not increase by a unit per layer. It is subject to a much more interesting and unexpected law.

You will easily understand this if you look at the summarizing drawing, Fig. 6. It contains a sequence of spheres tangential to each other's right sides. They belong to completed layers: the first, the second, and so on. These spheres are lined up along the lower horizontal axis, starting from the center of the first sphere (lower left), which is "fixed." These spheres are depicted as circles with the same diameter as the first sphere.

The drawing demonstrates the first three layers that are complete. They are shown by bold circles of larger diameter (actually, these are not circles, but spheres of larger diameter). However, this drawing can be infinitely extended to the right, allowing us to obtain as many layers as necessary.

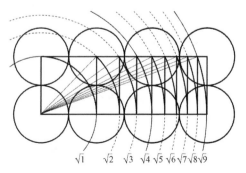

$\sqrt{1}$    $\sqrt{2}$   $\sqrt{3}$   $\sqrt{4}$ $\sqrt{5}$ $\sqrt{6}$ $\sqrt{7}$ $\sqrt{8}$ $\sqrt{9}$

Fig. 6. Spatial layers, dimensions, and transitions
between them.

41

I have placed the identical second line of spheres above the lower line to better understand the law of transitions between dimensions. This upper line of spheres is specific: while each lower sphere denotes the highest possible level of the corresponding layer, each upper sphere represents the lowest level of the next layer. That is, the upper sphere and the sphere directly under it belong to different but neighboring spatial layers. The horizontal line connecting the centers of the spheres in the upper line is distanced from the lower line by the diameter of the sphere. Now, you'll see exactly how this "magical" line will help us move between dimensions and how easily you can obtain the values of all projections and angles between them while transitioning.

Note that from this drawing we get our "magical" sequence of the space of spheres:

$$\sqrt{1};\sqrt{2};\sqrt{3};\sqrt{4};\sqrt{5};\sqrt{6};\sqrt{7};\sqrt{8};\sqrt{9}\dots$$

It's not just a beautiful mathematical sequence! Just think of the fact it reveals to us. If we pick two random spheres in any huge cluster of spheres, the distance between their centers will equal the square root of a whole number! It does not matter how many identical spheres you put together; you can take zillions of zillions or any number. This sequence is really magical, as is nature itself. Thus, in the optimal space of spheres, built on the basis of an equilateral triangle and a square, the following unique, simply stunning in its beauty, main Law of the Spheres is manifested:

*Within the optimal space of spheres of diameter of one, all distances between the centers of spheres composing it are equal to the square roots of whole numbers.*

This is the most important law—an incredibly magnificent and simple, simply "magical" law of nature, which underlies not only spaces, but also our consciousness. This is the very main Law of the Spheres that Thoth hinted at. This law, possessing divine beauty and harmony, comes from the very foundations of being. It was created exactly at the moment when spirit created the first sphere. The Pythagorean theorem and other laws of geometry are nothing but special cases of the main Law of Spheres.

Let's get back to Fig. 6. For example, let's take the second sphere from the lower line. It belongs to the first layer (zero dimension) of consciousness. The bold circle drawn through the centers of all subsequent spheres located along the lower line form nested layer spheres, with their radii being whole numbers divisible by the diameter of the initial sphere. Let us not forget that spatial layers (layers of consciousness) are located between the neighboring layer spheres.

Now let the real "magic" begin. Let's try to go through dimensions (frame a sequence of projections). For example, we can start from the third layer sphere, completing the third layer, and then move to the second one, completing the second layer. To do this, we must start moving upward along the circumference of the third layer corresponding to the square root of nine, from the point where it intersects with the lower horizontal axis, which is also the X-axis of coordinates. Having reached the point where the circle intersects with the upper horizontal line, we "abandon" the circle and "fall" vertically down, back to the lower axis. Now, look where we have fallen. This point lies on the surface of a sphere from the previous, lower dimension, while simultaneously being the square root of the dimensionality of the space we just left.

We must move up from this new point again along its circumference. Having reached the upper horizontal line, we "fall" vertically down again and so on. This way, we go through all the spheres. As you can see, you need to go through two additional levels to get from the second layer to the first one, that is, make three transitions. To get from the third layer to the second one, you need five transitions. Traveling from the fourth to the third one will take you seven transitions, and so on. The number of intermediate transitions increases by two at a time when the layer number increases by one. However, the distances and angles of transitions between neighboring dimensions decrease. You can obtain the angle values by connecting the center of the space (the center of the initial sphere) with the points where we've "bumped" into the upper horizontal line.

That's it. All laws can be derived from this simple drawing with no single linear segment that would not be equal to the square root of a whole number or to the difference of square roots. It also provides you with all distances between spheres of large and small tetrahedrons, an octahedron, all subsequent shapes, and distances to all spheres of all subsequent levels or layers, including all projections that we have obtained before or can obtain now.

You might then ask, what is there between spheres, where they do not touch each other? Is there emptiness? There is no emptiness in nature. You're right. Just press two balloons to each other or carefully explore the boundaries between soap bubbles stuck together, and you will immediately understand what exists between spheres. You will see that their contact pattern is not a point but a flat membrane. We have compared spatial cells to balloons or bubbles to better understand changing their shape from spheres to voluminous polyhedrons. However, the fact that you have pressed two balloons together doesn't prevent them from being balloons. Similarly, the spatial spheres remain spheres despite the fact they have composed cells shaped like spatial polyhedrons or prisms. Regardless of the shape acquired by a cell to fit in its place in space, there is always a sequence of unaffected values. These values are the distances between the centers of cells.

Let's look at the first day of creation, when the spirit created the second sphere, Fig. 7 (left). From this figure it is clearly seen that so far we have dealt with abstract mathematical spheres that only touched each other (in the figure they are shown by dotted lines). But real spheres are exactly twice as large!

When the spirit created the second sphere, a flat membrane appeared between the spheres, which passed exactly in the middle between the centers of the spheres—that is, divided the radii of both spheres in half. Connecting the intersection points and centers of the first two spheres with straight line segments will produce right-angled triangles with internal angles of 30 and 45 degrees. Plato believed that the entire world was created from these two types of triangles. But the most important thing is that if we connect all the main points inside the two intersecting spheres with straight line segments, Fig. 7 (right), then the values of these segments will give us a series of square roots of whole numbers from 1 to 9. This is amazing and incredible!

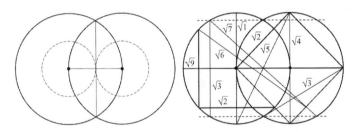

Fig. 7. The real spheres are twice the size of mathematical spheres (left). Distances between specific points within two linked spheres are also subject to the law of square roots of whole numbers (right).

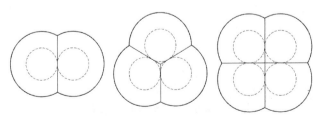

Fig. 8. The first three shapes formed by real spheres (straight lines correspond to plane membranes between linked spheres).

Two spheres make it possible to create the third and fourth, and also created the sequence of square roots of whole numbers, according to the laws of which the entire space is built. Therefore, the created first sphere already has the potential and the laws of creation of Reality and all the laws of geometry!

When the second sphere appeared, the first sphere just manifested the Law of Spheres hidden in it into the new space of two spheres. Adding new spheres, for example the third, fourth, Fig. 8, leads to the fact that this law manifests itself in larger and larger space—indefinitely!

Spheres have another extraordinary property. If we raise the "bar" (upper straight line) in Fig. 6 to a distance equal to two instead of one, we will be able to move through four dimensions at a time. If the distance is three, then we can move through nine dimensions, and so on. Drunvalo wrote that the ancient Egyptians could make spiral rounds with their consciousness by increasing steps. This particular property of spheres could be the basis for such "traveling."

If we apply the rules from Fig. 6 to actual spaces, we will be able to make transitions between dimensions (or levels of consciousness) by making spiral rounds, as shown in Fig. 9. The laws are the same. When we reach a new level, we shift the angle and then advance tangentially to the circumference of this level until we reach the next level, and so on. This way we can go through several levels and even layers in a single spiral turn.

The hashed line highlights the beginning of a smooth mathematical spiral we move along in the angled, broken line. Note that the turning angles are always subsiding. Additionally, they correspond to the angles from Fig. 6. Here is the sequence of angles:

90°;60°;45°;35.264°;30°;26.565°;24.095°;22.208°...

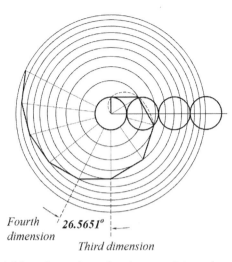

*Fourth dimension* / *26.5651°* |

*Third dimension*

Fig. 9. Spatial levels and angles in transitions between them.

We start from a 90-degree angle because the law of spiral angles extends below the first level we have used as the initial point for our space of spheres. It's all relative. The smallest sphere has its own level of consciousness, and we could start a countdown from it despite the fact it lies outside dimensions. Now imagine that the spirit Drunvalo wrote about ascended from the center of space and then moved through levels of consciousness. He needs to change his direction by 90° as soon as he reaches the surface of the initial sphere. Only then can he proceed further to the next level. After he reaches the first level, he makes a 60° turn and moves tangentially to the circumference of this level until he reaches the second level, and so on.

Levels belonging to the third and fourth dimensions are highlighted in the figure. You may see that we can move from the third to the fourth dimension by making a turn at an angle of 26.565 degrees. Memorize the value of this angle. We have seen it before, and we will see it again.

Spheres belonging to layers (the first, the second, and so on) have another amazing property. Indeed, nature has everything interconnected and everything envisioned. See for yourself: if we draw a tangential line to the large circle penetrating any sphere of the upper line in Fig. 6, instead of drawing a vertical downward line through its center, this tangent will divide the distance between the neighboring lower spheres by an integer number as it crosses the lower axis. This integer number is equal to the layer number of the lower sphere. In other words, it will cut a segment off the distance between centers of two neighboring lower spheres. This segment will fit this distance N times if N is the number of this layer. The tangent will leave a "notch" on the universal meter of space, as if saying, "Hey, look, this sphere to the left belongs to space number N." It's like the sphere itself has told you, "My number and my distance to the center are equal to N in universal meters. What else do you want to know about me? Do you want to learn the distances and angles to neighboring spheres? No problem, I have no secrets from you."

Thus, the tangent drawn from the center of the first upper sphere to the lower axis will mark an entire unit off our meter — that is, it will move exactly to the center of the lower sphere of the second layer. The tangent drawn from the center of the second (from the left) upper sphere will mark one-half off the distance between the centers of lower spheres of the second and third layers and move to the tangent point of the two lower spheres. When it is drawn from the center of the third sphere, it marks off one-third. If it starts at the center of the fourth sphere, it marks off one-quarter, and so on.

It took me some time to understand the practical purpose of this remarkable property of spheres deriving from the similarity of triangles (one should rather say it's the similarity of triangles that derives from the Laws of Spheres). By then, I had already learned that everything created by nature makes sense, and there is nothing that would not be connected with everything else in an absolute accurate manner. So I asked a question.

When you ask a question, the answer does not always come instantly. For example, once I had to attend a leadership workshop inside an American company. The topics of the workshop were of little interest to me, but I could not refuse to attend. Since I wasn't interested in them, I was half-listening while watching people and things both inside and outside the room. At some point, I noticed some excitement inside the room. The lecturer, a management specialist, was providing his audience insight into the methods of employee incentivization to help them work more efficiently. Having raised the problem of encouragement, he used a dog training method as an example. You know, when intellect develops its methods, it doesn't care who you are: a person, a dog, or a robot (it's even easier with robots). The lecturer asked the audience to pick one of two dogs that, in their opinion, would take orders most diligently: the one given a treat after executing each command, or the one given a treat on a case-by-case basis. Naturally, the audience picked the one given a treat on a case-by-case basis: it would not demand a reward, even if you forgot to give it a treat once again. The lecturer continued his lecture, satisfied with being understood. The audience followed him on his wave. I imagined two dogs. Indeed, I wondered which of them would be more diligent. Thus, the question was asked. Then, I've got a feeling that there was something equally wrong about the relationships-through-handouts in both cases, so I stopped thinking about it. Suddenly, a few seconds later, I obtained an answer: "The one which loves you more."

Next, it all happened in a similar way to my example with the dogs. I asked myself a question and forgot it. A few seconds later, the understanding came to me by itself. I felt admiration then, of the wisdom and perfection of the laws of nature, so perfectly combined with their simplicity and harmony.

Imagine you have suddenly found yourself in a space with unknown dimensionality. You have landed in a layer of spheres without any idea which one of them it is. If you don't find out the answer, you won't know where and how to proceed. Is there any way out? Sure, mathematicians will say it is necessary, for example, to define coordinates of three points of a large sphere, to restore the sphere using these points, to find out its diameter, and to divide it by the distances between the centers of the small spheres. This is the only way to define the number of a large sphere.

Here, we need to get rid of another illusion. If you ever find yourself in such a situation, your consciousness will be very different from what it is now; trust me. You will be unlikely to have a computer, a calculator, or even a ruler. It's actually good that you won't have any of these things. If nature's creation had depended on any technical stuff, it would not have lasted long. Can you imagine a creature from higher worlds solving integral or differential equations on a board? Or saying, "Wait a minute; I'll check the answer with a calculator" before answering your question. It would be funny. So, you will have no numbers or technical devices (there might be only simple whole numbers to form comparative categories). We use nine digits because we live in the space of the first nine levels of consciousness. Our mathematics is the mathematics of the first three layers of consciousness. Ancient civilizations went beyond the third layer. That is why they did not leave a single mathematical formula behind. All they had left for us were figurative drawings and geometric shapes. This is a kind of secret knowledge that does not need anything artificial to be understood and practiced. All you need is images.

You won't have numbers, but you'll have other capacities. They will help you work everything out without numbers. You will be perceiving the world differently. Perceptions are based on images and comparative categories corresponding to levels of your consciousness. You will be able to see or sense the presence of energy centers of small spheres and surfaces of large spheres composing space and points of intersections. You will be able to draw lines in your mind and compare segments.

There are lots of spheres at the bottom and lots of spheres at the top. How do you know exactly where you are? It seems impossible to find out without numbers. However, it turns out that the latter property of space makes it easy to find out. In fact, if you want to know the dimensionality of space you have landed in, you'll only need two right angles.

When the spirit was building space, he used only a voluminous cross (six beams) denoting directions and consisting of right angles and segments of sizes he chose as a base, as well as spheres produced by rotating these segments. He used clusters of energy to highlight spots on their surfaces, centers, and points of intersection. Scientists were even able to measure this energy in space. Therefore, you can measure everything in space with the help of the right angle. You don't need anything else!

Let's see how it works. You choose an energy center (that's how I'm going to call centers of spheres composing space) lying on the surface of the layer or the active shell,[2] which number you are going to figure out. You must then place the vertex of the right angle composed of two one-unit segments at the chosen energy center and direct one side (shoulder) toward the center of the space. Then, you must place the vertex of the second right angle at the top of the vertical segment of the first one and direct one of its segments toward the center of the space. In this case, the vertical segment of the second angle is like an indicator needle that will point on the horizontal segment of the first angle the layer number and the distance to the center. You will only need to visually compare two segments where one of them is a universal unit of space.

The mathematics of spheres is so excellent that it is involuntarily associated with fairy tales that give birth to fairy tale characters and fantastic adventures. Let's say that, from

Russian fairy tales, we know that once upon a time, a young man named Ivan traveled to the Three-Nine Kingdom located in the Three-Ten State, but no one has ever described details of his journey. We can do it now. Imagine, for example, such a fragment from a new fairy tale about Ivan.

"Ivan had already been walking for a while when he saw a huge mountain, round and smooth like a ball. It was so big that one couldn't get around or over it. Ivan recalled what the wizard had told him. There should be a door in the mountain. Only the chosen one could find this door, and those who find it will fail to open it without help. Ivan asked the mountain, having put a bold face on:

"Dear Mountain, you're as round and beautiful as a ball, could you show me where the secret door is? Please, open the path to the Three-Nine Kingdom."

"What's there for you in the Three-Nine Kingdom?" the mountain replied with such thunder that made the Earth shake.

"Stay here or go home."

Ivan wasn't scared.

"I can't go home. Koschei's hordes have tormented my land. I want to save my people, to free them from Koschei," he replied.

"There is a door in front of you," the mountain rumbled. "It can be opened by the one who has the key-to-all-doors."

Ivan scratched his head but then realized he had no other choice. Nobody had told him about the key. He had no key. He had nothing at all except for pants, a shirt, and empty pockets. The young man became distressed. He was going to return home empty-handed when he recalled what the wizard had said: "If you are unable to solve the puzzle, just say, "As much up as down, as much left as right, as much forward as backward."

Be that as it may, Ivan thought. Then he repeated the words just like the wizard had taught him.

Suddenly, the mountain trembled. A stone beamed with wondrous light and started sliding aside right in front of Ivan. A magic door opened.

What a miracle! Ivan thought in delight. He stepped into a huge doorway and saw two wide staircases made of glowing stones resembling crystals: one led to the top, and the other spiraled downward. They shimmered with different colors and sparkled like stars. The steps were tricky and very difficult to climb. These were no ordinary steps; they were alive. Each one sang its song. One could see neither where they headed nor where they ended. Ivan didn't know where to go. He saw a huge tree far overhead, with a pure stream flowing from under its roots. Its waters were flowing down the steps. It felt like water, but Ivan's hands didn't get wet. It looked like another miracle.

"Help me," Ivan pleaded, "Dear Mountain, help me find the right way."

"Ask your questions," the mountain replied, "I have no secrets from the one who has the key-to-all-doors."

"Tell me," Ivan said, "where the stairs are going, where they begin, and where they end, and how far is the Three-Nine Kingdom from here?"

"The beginning is at the bottom; the stairs are endless," the mountain replied. "If you go down, you'll find a short cut home. If you go up, you'll find what you're looking for. Count the steps, so as not to be mistaken. You'll meet my sisters on your way. They'll help you because you have the key. I am the fourth of them. The higher you go, the easier it will be to climb."

"Tell me how many steps there are to your third sister?" Ivan asked, ready to memorize every single word the mountain uttered.

"Twice in fours."

"How far is your fifth sister?"

"Twice in fives."

"How far is the beginning of the stairs?"

"Four times in fours," the mountain replied. "It takes as many times as many steps from the beginning of the stairs to reach each sister as each sister is in succession. You won't be mistaken."

"Thank you, dear mountain," Ivan said as he began climbing the stairs. The way back will be easier, he thought.

As soon as he approached the first step, he heard the sound of wings fluttering overhead. It was like the wind had fallen from above. A huge firebird landed in front of him. Ivan had heard about them, though he'd never seen one. It was three times the size of Ivan. No one knew where and how it was born. Its eyes were glowing.

"Hello, Ivan," the bird said. "You have helped others, so now it's your turn to be helped. Get on my back, and I'll take you to the Three-Nine Kingdom in three-nine faster time."

"Thank you," Ivan replied.

He climbed on the magic bird's back and clung to it as tight as he could. The firebird swung its wings once, and the land below seemed to disappear. Ivan felt he could barely breathe. He clung to the bird and seemed to have lost his ability to move. It swung its wings for the second time, and it was even more breathtaking. The stars appeared all around them. The bird swung its wings for the third time, and it was the end of their journey. It landed on a field with flowers unseen before. Ivan jumped down and couldn't utter a word because of his excitement. It felt like he'd forgotten all words at once.

"Here is the Three-Nine Kingdom," the firebird said. "Here are the stairs. You know how to get back."

Ivan looked around and saw an even steeper mountain standing there, in the middle of the sea, with a castle on its peak, shining as if made of precious stones, all glowing. It was so high that clouds obscured the towers. You couldn't see where they ended. A staircase, like a crystal rainbow, led to the castle, and its steps were low. The sea was swelling under it. It was no ordinary sea; it was crystal clear, though its bottom was not discernible. It was filled with the water of life. A song was singing in the air, so beautiful that it was impossible to become distracted, while his feet were walking up by themselves. Ivan stepped on a crystal bridge, but suddenly his heart trembled: he recalled why he'd come here in the first place. People's voices sounded in his ears. His people were suffering, moaning from Koschei's torments, asking for help.

"Call me if you need me," the bird said. With these words, it flew up and suddenly disappeared.

Ivan was left alone. He looked around and saw...

You can continue this fairy tale yourself.

Lest we be reproached for being too attached to Russian fairy tales, let's recall, say, Arabian ones. Let's recall a phrase from the story of Ali Baba and the Forty Robbers: "Sim-Sim, open the door!" It made the door open. Let's find out what kind of a universal magic key it was, a key that could open magic doors. If we add "Sim" to a universal meter of space we are in, we will get Sim-Metry. However, we have "Sim-Sim" here, which is a dual symmetry with the cross as its symbol. So, we've got a universal key. Once it is attached to a sphere, the door to another dimension will open.

I would not advise opening doors to lower worlds; you may let evil spirits in. It has already been done by unscrupulous magicians longing for power. It was for this reason that ancient Egyptians believed that the one craving for power is, in fact, a slave. Eventually, they did not receive power but instead became eternal slaves to the dimension they had brought into this world. Remember, as the djinn used to say, "I am a slave of the lamp." He meant to say he was a slave of a closed, limited space. Imagine, being a slave, forever! It's difficult to come up with a fairer punishment for the thirst for power over others. At the same time, it is difficult to find a more terrible punishment.

If you look at the relationship between spheres in a slightly different way, you can see that the square of the circumference (or sphere) radius allows us to determine its number within a sequence of spheres. This is the squaring of the circle people have been failing to find for thousands of years.

Now, as we have learned about the structure of the space of consciousness, which "defines our being," we can finally understand the actual meaning of figurative comparisons and allegories associated with this issue. They have come to us from religions, epics, and fragments of ancient knowledge. Thanks to the map of dimensions and transitions between them, we can distinguish right information about space from false or distorted representations people have accumulated over many centuries.

For example, Dante described hell as consisting of nine levels. Most likely, he had adopted this gradation from more ancient knowledge. Dante was a poet and a man who genuinely loved a single woman for his entire life. Although the woman passed away at a young age, this love allowed Dante to raise himself above ordinary life. Dante was neither a mathematician nor a philosopher. That's why he can be excused for some imprecision in describing spheres and dimensions. He didn't claim to be scientific anyway. However, his nine levels of hell appeared for a reason. Dante didn't invent that number. Calculate how many levels there are in the first three layers of space of consciousness. You've already learned what the nine levels of hell mean, namely: $1 + 3 + 5$. Surely, the word "hell" can be used only figuratively here. It is the state of consciousness corresponding to one of its first nine levels or the first three layers. Now count how many levels there are in the fourth layer. You'll get the number seven. They correspond to the Seven Divine Spheres, or seven levels of Purgatory (if you take everything that is above the fourth layer as heaven).

Humanity has lost its comprehensive knowledge of the structure of levels of consciousness. Most ancient works, as well as their adherents, were destroyed by the church. However, some fragments of the knowledge we inherited from the Egyptians or even more ancient civilizations have survived to modern times. They have been passed down from generation to generation (this was how Dante learned it), inspiring scientists to seek the truth. As an example, we can view an image from Robert Fludd's book (XVII century) depicted in Fig. 10.

Fig. 10. A picture from an ancient manuscript. A stairway to heaven resting on the previous three layers of consciousness.

You might say, "That's nonsense! I'm not going to believe it! I don't want to live in hell. Hell is what's underneath. Heaven awaits us at the top, as they say in the church." Well, then please answer a single question, dear reader. In the church, you're told that God once banished humanity from heaven. If he banished them from heaven, where did humanity end up? Who has told you that there was a flying island called Earth between heaven and hell?

If I were an artist, I would depict all nine levels of hell on a canvas. The nine levels of hell are the nine levels of consciousness. First, I would place them within right proportions, like the floors of a building on a draft, with stairs between them going at certain angles. Then, I would draw scenes and actions typical for each floor as they emotionally correspond to their level. Thus, I would draw all the torments humanity has to endure on the fifth, fourth, and third floors, because these are the main levels of hell. I would depict people burned in fires, murdered both during wars and ordinary life, and blown to pieces by nuclear explosions. I would depict people losing their loved ones forever and suffering from diseases, slavery, violence, deceit, betrayal, heavy and numbing work, hunger, cold, and lack of love. Look at this canvas. Are you still going to say you don't live in hell? What would you call this place then?

Dear reader, I don't know how you got to this layer — whether it was a coincidence, the will of chance, or a conscious choice. This information is available to your projections from higher dimensions. After all, your projections exist in all dimensions at once. If they weren't there, you wouldn't be here. You've just become very fascinated with this projection in the second layer of space of consciousness and decided to linger here for some reason. It vibrates in resonance with your consciousness, making you see the world in its colors and sense it with its senses. Your projections call you from higher worlds, with the call of higher spheres.

Before you resent God for creating the hell you live in, you must understand a very important thing: God did not create hell. There is actually no single individual god. An individual almighty "God" was invented by churchmen. Hell was created by humanity due to their imperfect consciousness. It appeared through erroneous programs adopted by the intellect, as well as through distorted relations to one another and to the entire outside world. Such programs were artificially intensified and imposed upon humanity by those wishing to destroy it. Humanity could easily end the hell and live in a beautiful, nature-created happy world if they evolve their minds enough to realize the real causes and consequences of their unenviable position. However, many of us prefer to stay hibernating in our ignorance, proceeding further by trial and error and experiencing incredible torments of real existential hell (since we can't call it life) in a condition of complete lovelessness.

Considering the structure of spaces, no one can promise you that after death all people will move to the "fourth dimension", which in our drawing is in the third layer, or, even more so, to the "seventh heaven", corresponding to the fourth layer, because your heart will be weighed on the scales of Maat. Only those whose hearts weigh no more than the truth will get to the higher worlds. Some of us might get to higher spheres indeed. If you are an ordinary person trying to live by your intellect, like most people, you might become careless over the course of your life. In this case, when death captures your consciousness, with metamorphoses of changing shapes and scenes, you might be reflexively captured by new scenes, as you used to do all your life. Your consciousness will be immediately drawn into a vortex that will swirl you and throw you to the wrong island. You may find yourself in lower dimensions of the second layer of consciousness. These worlds will have lower vibrations and folded levels of consciousness, with fewer dimensions and much more suffering.

*The violent occupy*
*All the first circle; and because to force*
*Three persons are obnoxious, in three rounds*
*Each within other sep'rate is it fram'd*
**Dante Alighieri**, The Divine Comedy, Vol. 1 (Inferno)

I don't know how Dante learned this information, but he told us the gospel truth here. This circle or layer of human levels of consciousness comprises three levels, one below another. Most people are at one of those three levels of consciousness. Therefore, a person has a great chance to get to "flat" and "linear" dimensions instead of higher spheres. It depends on one's condition. There will be no way out of there because one might not remember the need to apply, in one's consciousness, to higher dimensions to find a way out. Nothing will happen without a well-conceived desire, proper strength, aspiration of spirit, and accumulated potential. It is the well-conceived and spirit-backed intention or call that helps you generate a single vector from faith, hope, and love. This vector can take you to the worlds where you deserve to be. You do not need to memorize values of all angles to move between dimensions. Your consciousness contains this information; thus, it can find the right way itself.

I have shown you this picture so that you can get an initial idea of spheres in the space of consciousness and realize where you stand at the moment. You must also realize that you can transition between dimensions of space only at certain angles, with the main angle equal to 90 degrees. Perhaps you can make use of the symbol of the cross if your consciousness is able to remember it after death. The cross is made up of right angles only. However, when you leave the three-dimensional world and abandon your body along with its mathematical and logical apparatuses, you will hardly remember that there are degrees and angles at all.

You must evolve a sense of constant "vertical" connection with the Superior Mind of the Universe or with your Higher Self in the course of life. It means that you must fill every single moment of your life with this seamless connection along an upward line, starting from your heart and traveling at an angle it will choose itself. In this case, such a connection will reflexively help you out after your body dies. The only condition is that this connection should be generated in your heart by senses, not by intellect. If you recall and use it at some particular moment, you will be able to instantly snatch your consciousness from the linear world of an infinite sequence of shapes and scenes and move to a much better world of higher dimensions.

After you close this book, I advise you to read some Russian fairytales about Ivan. Do you remember how they often begin? "The father had three sons: two of them were smart, and the third one was a fool." That is, two elder brothers were always busy; they were working, trading, doing their own business. Ivan was resting on the stove most of his time. In general, he was not smart at all, and it is not clear what he was thinking of. Nevertheless, when it was necessary to go to the Three-Nine Kingdom, Ivan braced himself and went there without hesitation.

It turns out that the Three-Nine Kingdom really exists. Fairytales don't lie to us. It is located exactly in the Three-Ten State. However, an ordinary person won't find it. One can travel around the entire world without finding it. Ivan, however, found the way and, most importantly, reached the kingdom! This is because he wasn't ordinary. What kind of person was Ivan? You will understand yourself if you follow his path described in our fairytale. You have enough knowledge for this now. All you need is a piece of paper, a compass, and a right angle. You will also understand why an ordinary person will not find it.

You will need the same tools God used to create the universe, as according to a painting by an eighteenth-century artist, William Blake. In this painting, God also had a compass and right angle, Fig. 11. However, if you are fascinated with studying nature's creations, try not to be like Newton, whom Blake depicted on another canvas, Fig. 12. The famous scientist was also holding a compass. At that time, this simple tool was traditionally considered a symbol of God. Pay attention to the last painting. You can see that the brightness of colors and light emanated by nature fades from left to right, gradually thickening into impenetrable darkness. Blake believed that a narrow-minded and lopsided view of the world, typical of Newton and his fellow scientists, was tantamount to blindness. Studying only the physical aspect of the world while ignoring its other aspects doesn't bring any light. On the contrary, it pushes this world into darkness.

Fig. 11. God as the architect of the world on William Blake's canvas.

Fig. 12. Another painting by William Blake: Newton with a compass in his hand.

## Endnotes

1.  In fact, the point in the middle of the segment is the very starting point that space originates from. This becomes clear when you remember how spirit created the first space of spheres around itself. However, spirit did much more than just create space. It powered all its centers, surfaces, and points of intersections between spheres with additional energy. Scientists were even able to measure this energy in space. There is also another essential point allowing you to obtain a better understanding of consciousness' space. When the second sphere appears after the first one, it denotes more than just the appearance of the second new object (sphere). Two spheres together become a new object. It is this new object that continues expanding when you add new spheres. Then another new object consisting of three spheres appears. Next, the objects contain four, five, or more spheres. Each new object preserves all the peculiarities of the objects embedded in it and acquires new peculiarities previous objects did not have. As we will see later, these peculiarities are associated with new dimensions or new levels in the space. According to a vague analogy, the hivemind of human society includes the minds of individuals while evolving some new peculiarities individuals don't have. However, this statement does not apply to the intellect. Osho once said that the intellect of an individual idiot is much more conscious than the collective intellect.

2.  Each living being has an active shell around it sized as its level of consciousness. This applies equally to humanity, to the planet, and to the entire galaxy. Real life is much more interesting than any fiction. By defining (or measuring) the diameter of the outer active shell of the galaxy, you can determine the consciousness level of life in it. You don't even have to fly there, land on its planets, take samples, explore creatures, and so on. Can science fiction writers imagine anything like that!?

**Chapter 3**

# The Ancient Mystery of the Flower of Life

Exploring the space of spheres, spatial layers, and shapes is an interesting and exciting activity that allows for an understanding of many things. All laws of geometry are derived from the Laws of Spheres, which is wonderful and amazing. When you dive into studying the Laws of Spheres, you dive into the flawless spatial harmony. It is difficult to compare this sense of being related to something so perfect with anything in this world. My soul, no matter how it was disturbed by the disharmonious world we live in, was filled with peace and harmony in no time after I started exploring true sacred geometry originating from nature itself. It's disheartening to see what people have turned this wonderful and amazing world into. One of our main goals is to bring original harmony and beauty back to this world. I've had a particularly strong sense of harmony since experimenting with light and sound, but we will talk about that later. In the meantime, let's get back to the space of spheres.

There are several possible ways of framing layers within the optimal space of spheres. The view of space depends on the initial point we take for its center. More precisely, it depends on the simplest shape (Fig. 4) we will take for the center of space. The only exception is when we deal with a pentagon and the shapes derived from it. The pentagon case is quite peculiar, but we will discuss it later.

There are only three ways of building an optimal space. At least I have managed to find only three of them. Which way is the right one? Might all of them be the right ones? If all of them had been the right ones, there would have been three different methods of creating a sapient world. Moreover, sapient beings inhabiting these worlds would have also differed not only in their intelligence level but also in terms of space type. Because there's no such thing in nature (at least, I've never heard of anything like that), only one way is the right one. To understand which way is the right one, we need to apply each of them and evaluate the results. So, let's remember once again that in the optimal space of spheres each sphere touches neighboring ones. The distances between the centers of all spheres are the shortest possible and equal to the square roots of whole numbers.

Such a space can be built around a single initial sphere if you choose it for the center of the space. First, imagine you have put a sphere on the table. This sphere can be surrounded by six identical spheres on the plane of the table, so each sphere will be in contact with the neighboring ones and with the central sphere, Fig. 13 (upper left). It's easy to imagine. From this point onward, I've placed photos of various shapes made of glass balls stuck together. These images will help you better understand and visualize the resulting spatial shapes composed of spheres.

Fig. 13. The sequence of constructing shapes of the optimal space around a single sphere. The drawings correspond to the first (upper left) to fourth (lower right) steps of construction.

Next, we need to shift from a two-dimensional image (the traditional view) to a three-dimensional one. When building layers, you should always remember that new spheres "seek" to occupy the innermost cluster point as they are being added on the surface of the already existing shape from the spheres. In other words, they strive to take the closest possible position to the center of the space. This is a general rule of thumb; otherwise, it is impossible to achieve an optimal space. This way, we can easily add two triangles composed of spheres to a hexagon we have already shaped. Triangles should be added to the opposite sides of its plane, Fig. 13 (upper right).

The resulting shape consists of four hexagons intersecting a common center at the angles of 60 degrees. So far, so good. Thus, one sphere can be surrounded by twelve tangential spheres. Twelve is quite a celebrity among numbers, isn't it? It usually means something complete or holistic, like twelve hours, twelve months, twelve apostles...

It is of no surprise that some scientists have been caught by mathematical features of this shape, which Buckminster Fuller called a cuboctahedron or vector equilibrium. For example, a popular new age physicist, Nassim Haramein [2], suggested that this shape was the Flower of Life itself. He suggested that the vector equilibrium underlined the structure of the world. Based on this idea, he developed the "unified hyperdimensional theory of matter and energy" or the theory of the "Holofractographic Universe." A bit later, you will see that the concept underlying this theory, and its mathematical calculations, is inherently erroneous. This theory takes us away from understanding reality rather than bringing us closer to it. Let's not get distracted and proceed with building our space.

Using the same rule, we can add a few more spheres to the resulting shape to obtain a large octahedron, Fig. 13 (lower left).

Then, we can get the shape depicted in Fig. 13 (lower right) by adding eight more triangles to this octahedron. These triangles are clearly discernible in the photo. By adding a few more spheres, we can obtain a shape from Fig. 14 (left), followed by a prism from Fig. 14 (right), and so on. If we mentally take segments of straight lines and connect the centers of all tangential spheres on the surface of the resulting shape, we can obtain a prism of a certain shape.

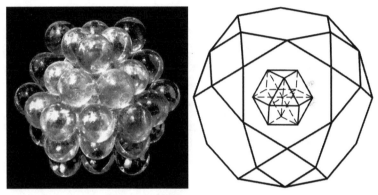

Fig. 14. The fifth (left) and the sixth (right) steps of construction of shapes of the optimal space around a single sphere.

The shapes in Fig. 14 conceal a lot of interesting surprises, as well as all previous shapes. If we mentally slit either of them through the center along the plane of any of the four inner intersecting hexagons, we can see the spatial structure of spheres in the section as in Fig. 15. This figure clearly demonstrates that, within this shape section, spheres compose a dense optimal structure consisting of a mixture of equilateral triangles and squares. You can continue this structure indefinitely, but it will still consist of equilateral triangles and squares only. You won't find any other type of shapes in it.

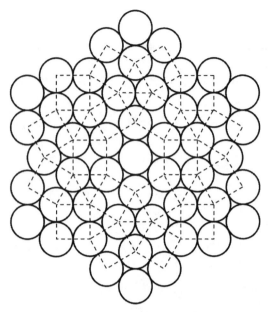

Fig. 15. This structure consisting only of equilateral triangles and squares can be obtained if we mentally slice the shapes from Fig. 15 along the plane of any of the four central hexagons.

We're not done with surprises yet. Besides this section, there are two other sections in the shapes from Fig. 14 that do not pass through the center. One type goes through triangles of the shape shown in Fig 13 (upper right), and another one goes through squares. These two types of planes of the section are situated at very specific angles to each other. In one type of plane, spheres are shaped like equilateral triangles only (Fig. 16), and in another one, they are shaped like squares only (Fig. 17). It can be assumed that the properties of the actual space of spheres will be different within each of the three types of planes of the section.

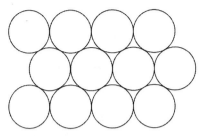

Fig. 16. In the cut planes passing through the triangular faces of the prism in Fig. 14 (upper right), the spheres form a structure consisting only of equilateral triangles, as shown in this drawing.

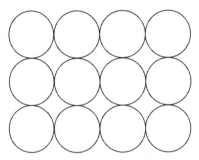

Fig. 17. This structure, consisting only of squares, can be obtained if we cut the prism from Fig. 14 (upper right) along the cut planes passing through its square faces.

The optimal space of spheres has another unique and amazing property. It has all other spheres surrounding each of its componential spheres situated on the surfaces of the spheres (Fig. 18) with diameters subject to the rule of square roots of whole numbers. The entire space is penetrated with spheres and composed of spheres only. Within the optimal space of spheres, there are spheres everywhere! Isn't that magic?

Thinking all these layers and shapes offer great value, such as space construction, is wrong. "What's wrong?" you might wonder. "Is such space construction impossible?" It's possible. We have just seen it. "Aren't the distances between the centers of all spheres equal to the square roots of whole numbers?"

They are. "Does such space exist in nature?" It does. "So what's wrong then?"

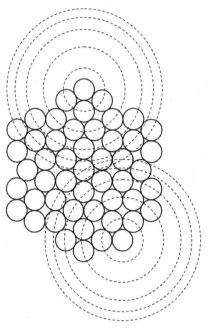

Fig. 18. Another "magical" property of the optimal space of spheres. If you look at any sphere within this space, you will notice that other spheres around it are located on the surfaces of spheres of different diameters, which are subject to the same rule of square roots.

Wrong is our way of thinking that is selfishly used to always put someone or one thing in the center of attention. We rely on false stereotypes carved into our brains back in school when framing such space construction. If you try to build this space yourself, you'll notice that it is imperfect in as far as the second step goes because it contains uncertainties. If you build it by adding spheres step by step according to the rule on the shortest distance to the center of the shape, then you can reach a dead-end at certain spots.

For example, as you might notice from Fig. 13 (upper right), the upper and lower triangles (above and below the hexagon) are turned by 60 degrees relative to each other within the vector equilibrium. If you don't turn them, you won't be able to build space any further. Think for yourself, how should a new sphere "know" that it is necessary to occupy the vertex of the triangle? After all, the vertices are equally distanced from the center. Which one of them has been actually turned, and which one hasn't? There are no criteria because the hexagon is symmetrical. Thus, we should conditionally add whole triangles (the first condition) to the hexagon, and they should be turned relative to each other (the second condition). Something is wrong here, however. This does not fit within the concept of perfection.

The next conventionality will not take long. We need to add triangles to the shape in Fig. 13 (lower left) again. What's more, there should be several of them at once. Similar conditions appear again and again. This space is not perfect; it is impossible to build a space of consciousness based on it. Following the logic of our intellect, we have made a mistake. We have completely forgotten about the evolution of shapes because they are not static. We have forgotten an important fact. When, let's say, two spheres are combined, they become a new individual object called "two spheres." When another identical object is added to the "two spheres," four spheres become a new object called a "tetrahedron" because it is the optimal shape for four spheres. Now, space should be built around this new object instead of a single sphere. After all, when you live in your house, it is always a single house for you, no matter how many rooms it has: one, five, or ten. If you live in a two-bedroom house and then build an extra bedroom, it will still be your house, just larger. It will only have more dimensions if we consider each bedroom as a separate dimension.

We have just learned a practical lesson on how easy it is to walk on the wrong path when following the intellect. As a result, we've got space with someone in its center; space is built around someone with a lot of various conditions to be observed simultaneously. Everything intellect does is overburdening itself with conditions. The number of conditions and dogmas constantly increases due to its activity. For this reason, modern science applies to an increasing number of conventions and postulates. Therefore, given the outstanding characteristics of space, we can conditionally call this space "space of an egoist" to memorize this important lesson. On the other hand, this name might not be conditional at all.

Sure, the name we have given to this space refers to our idea of space, to our intellect and our way of perceiving reality, rather than to space itself. Space itself, of course, is not to blame for anything. We have just chosen an incorrect center. When someone appears in the center, one facilitates generating a false center of perception with a distorted view of the world being framed around it.

Since we have some issues with building an optimal space around a single sphere, we will try to build space around the next shape, a tetrahedron. In this case, we will obtain a set of levels and layers with their shapes (Fig. 19) utterly different from the layer shapes of the previous case. However, the law of square roots should also be implemented here with impeccable accuracy.

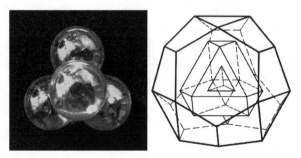

Fig. 19. Three-dimensional shapes of different levels of optimal space built around a small tetrahedron (in the center).

Although new space has improved qualities, it is also imperfect and requires some conditions to be fulfilled. It also provides us with a distorted view of the world. For figurative comparison, this space can be called "space of a religious person." Judge for yourself: we have built this space around the simplest shape consisting of four elements. We had to build it while observing a lot of conventions to make it look good. If your consciousness evolves a view of the world centered around a personified god with a name consisting of four letters, for example, then it must be based on many conventions. Such conventions are religious dogmas; it is them that make such a view of the world look coherent. You can't develop a religion without dogmas. Do you realize from this description who the real author of religion is?

We have made the same mistake by following our intellect. Guess what the trick is? Again, we haven't considered evolution, that is, the transformation of shapes. A tetrahedron consisting of two pairs of spheres is the optimal shape, but only if you're dealing with four spheres. If we add two new spheres to the tetrahedron, it will automatically transform itself (under the influence of attractive interaction force between spheres) into the shape optimal for six spheres. It will be transformed into an octahedron, Fig. 20 (upper left), a sustainable, symmetrical, and optimal shape for six spheres. Thus, we can say that an octahedron consists of three pairs of spheres. Such a "trinity" gives completely new properties to space. The octahedron has no more space to transform. Space is being built around it in a perfect way. It's like magic, with no need for any conditions to be fulfilled. Thus, we get a really peculiar, perfect view of space our minds are built upon. Here, the result will be completely different. The new spheres we have added will accurately take strictly allotted places according to the rule of the shortest distance to the center, building appropriate spatial levels and layers.

Fig. 20. By adding new spheres, the octahedron (upper left) is turned into a large tetrahedron (upper right), and then into the star tetrahedron (lower left) corresponding to human consciousness. By adding 24 more spheres, the star tetrahedron will be transformed into the figure shown in the lower right image, the projection of which on a plane will give you the ancient symbol of the Flower of Life.

So, when we build the first spatial layer around the octahedron (it will be the second layer in relation to the initial sphere) and add four new spheres to it, we get an already familiar shape, that is, a large tetrahedron, Fig. 20 (upper right). As we already know, this belongs to the completed second layer (see Fig. 6) because its largest possible size is equal to the square root of four or two. After adding four more spheres, we can turn the large tetrahedron into the most famous "star tetrahedron" (see Fig. 4 and Fig. 20 (lower left)), a spatial shape corresponding to human consciousness.

When we talk about spatial shapes, levels, and completed spatial layers, we need to remember the point we start measuring these levels and layers from. Again, we need to get away from the logic of intellect to avoid falling into the same trap for the third time. Thoth's tip helps to find the answer to the question of how to define the dimensionality of spatial shapes. Thoth said that people on Earth are between the fifth and sixth levels of consciousness. Knowing that this consciousness corresponds to the star tetrahedron and counting the levels starting from the first sphere, we get an answer. The answer has been very simple and natural, as everything in the space of spheres is.

All we need is to follow the rule we learned in the first two chapters: each spatial shape composed of spheres belongs to the levels of space that can be transitioned within this shape without breaking the law of connectedness of spaces. If you don't follow this rule, you can easily get misled.

For example, the longest possible distance between the opposite vertices of the star tetrahedron is equal to the square root of six, but despite this, the shape does not belong to the sixth level. It belongs to the fourth level. This is easy to understand when you remember that the star tetrahedron is framed by two large tetrahedrons inserted in each other, with both of them belonging to the fourth level. The star tetrahedron only completes this fourth level within spatial symmetry. Although it is impossible to transition to the sixth level (since there is no fifth level in this shape), this level still actually exists and vibrates. I cannot say for sure how it affects human consciousness. Perhaps this higher level sets the driving force that prompts human beings to proceed with their mind evolution to speed up the filling of the missing fifth level. In any case, the star tetrahedron is an unbalanced shape in terms of spatial dimensions. It is an intermediate, transitional shape situated on the way toward building a third layer of consciousness. Therefore, this state of consciousness experienced by a person is not sustainable and not harmonious, as noted by Drunvalo (while he was just repeating Thoth's words). However, other, higher levels of consciousness cannot be built, skipping this stage. Because of this inharmoniousness, human beings are subject to destroying and experiencing numerous sufferings caused by their imperfect consciousness, leading to erroneous decisions and actions.

Everything is more straightforward with all other shapes because they are completely balanced. The highest possible level matches their largest possible size. Their smallest possible size corresponds to the level with its number by one higher than the highest possible level of the shape of the lower dimension embedded in it. Let's remember that each shape contains all other lower levels belonging to the shapes it has originated from. Therefore, to determine the highest level this shape belongs to, you just need to measure its largest possible size. Starting from the largest possible size, it is possible to graphically frame a continuous sequence of interconnected projections and transitions between all embedded shapes and descending levels until we get to the lowest level along this descending angled line. The lowest level will be equal to the square root of one.

We've already seen how this rule is applied in the example of an octahedron embedded in a large tetrahedron. The octahedron has the lowest level equal to the square root of one, and its largest level is equal to the square root of two. Therefore, if we look at Fig. 6, we will see that the octahedron is a shape of the second level of consciousness, and, at the same time, it belongs to the first level of the second layer.

What will happen when the missing level in the star tetrahedron is filled? Something very interesting. Probably you will be curious about looking at the shapes of consciousness levels that are higher than the consciousness of an average person. We will switch between levels, one by one. If new spheres are gradually being added to the star tetrahedron, the levels a human being is missing to complete the third layer and acquire harmonious consciousness will be filled.

When we add 24 spheres to the star tetrahedron (hexagons composed of spheres should be placed around each vertex, or squares should be placed around each face of the cube, which is the same), we will obtain a completely symmetrical shape, as depicted in Fig. 20 (lower right) and Fig. 21. It is a prism, with its faces being hexagons and squares; its form is much closer to the sphere. There is not much we can say about the shape of complete human consciousness, which corresponds to the consciousness structure of a human being who completed the planetary level of one's evolution. Such a human being can transition to the cosmic level. If we draw a projection of this shape, we will get what is called the Flower of Life, Fig. 22. You have just learned the ancient mystery of the Flower of Life's symbol, including the key to its understanding, which was lost in the depth of millennia.

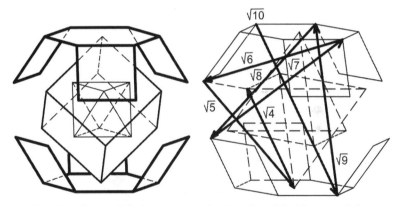

Fig. 21. A spatial shape completing the third layer of the optimal space of spheres is built around an octahedron. The inner cube is formed by the vertices of the star tetrahedron corresponding to the consciousness of a modern human. A sequence of transitions from the highest, the ninth, to the third dimension within the conditions of space connectedness is shown on the right.

Fig. 22. Revealing the ancient mystery of the Flower of Life.
This symbol corresponds to the shape of the completed
planetary level of human consciousness. It is a fully balanced
consciousness. By reaching it, a human being becomes truly
conscious and proceeds to the path of development of cosmic
levels of consciousness.

The Flower of Life symbol can be found in ancient temples
and objects everywhere in the world, including Egypt. I have
managed to find the symbol of the Seed of Life (which is an
octahedron) even on ancient objects and walls of ancient
Armenian temples, Fig. 23 and Fig 24.

Fig. 23. The ancient symbol of the Seed of Life found in
Armenia on the wall of an ancient temple.

Fig. 24. The ancient symbol of the Seed of Life found in Armenia on an ancient object (top), where it is depicted next to the merkaba symbol. Curiously, the ancient Slavic swastika (bottom) ended up in the same museum. I am glad that in Armenia, there is still something that is strongly forbidden and often destroyed in Russia to make people forget their great past.

Unfortunately, humanity completely forgot the meaning of these symbols several millennia ago. Only now have we revealed this mystery again. We have witnessed the Seed of Life's entire path, from its "growth" to becoming the Sprout of Life to the Flower of Life appearing from the sprout. It is also natural because it happens in nature. To improve visual perception, not all the possible lines connecting all neighboring spheres' centers are depicted in Fig. 21 and other figures. Only some of them are shown to provide you with a general spatial visualization. The new shape has no "sharp" vertices, as there are in the star tetrahedron. Also, it is fully balanced in terms of levels and transitions between them. Fig. 22 (right) shows a sequence of projections and transitions between the levels within the shape, starting from its highest, tenth, level that is equal to the square root of ten (ninth dimension), gradually descending to the star tetrahedron embedded in it. You already know how to expand this sequence of projections within the star tetrahedron up to the first level. Since the largest possible size of this shape is equal to the square root of ten, it completes the third layer and reaches the first level of the fourth layer—that is, the first level of the Seven Divine Spheres.

When a person masters all nine consciousness levels and passes to the fourth layer, the shape corresponding to this consciousness takes exactly the form shown in Fig. 21 and Fig. 22. Only by mastering one's consciousness can human beings reach the levels associated with the Seven Heavens, namely the seven levels of the completed fourth layer. However, the shape in Fig. 21 and Fig. 22 only leads to the first of the Seven Heavens.

Let's see what happens next. If you continue adding spheres to this shape, following the rule of consecutive filling of levels, it will be transformed into a unique shape after the second or third step. It's unique because it consists of four shapes, as shown in Fig. 25. To see each of them, you just need to highlight certain lines connecting its vertices.

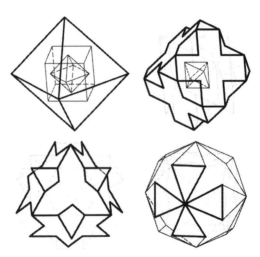

Fig. 25. Shapes of the completed fourth layer of consciousness, corresponding to the Seven Divine Spheres, reveal the mystery of the three sacred symbols.

This extraordinary spatial shape consists of a large octahedron, Fig. 25 (upper left), Greek crosses, Fig. 25 (upper right), six-pointed stars, Fig. 25 (lower left), or Maltese crosses, Fig. 25 (lower right), depending on what you pay attention to. All four shapes correspond to the eighteenth level of consciousness (square root of 18) — in other words, the second level of the fifth layer. It is also a wholly balanced shape. If you wish, you will be able to frame an entire seamless sequence of projections and transitions, ranging from the highest, eighteenth, level to the tenth level (square root of 10) belonging to the shape of the third layer we have just discussed above. Aren't they harmonious shapes? These are not only curious pictures. You have just learned the mystery of the origin of the three sacred symbols: Greek cross, six-pointed star, and Maltese cross — a mystery that has also been concealed in the depths of millennia. As you can see, these symbols were not invented by someone once; they were born of nature itself. Fig. 26 shows two shapes of the 22nd and the 30th levels of consciousness. The Tree of Life continues its growth.

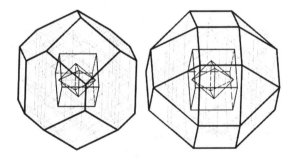

Fig. 26. Shapes of the 22nd (left) and 30th (right) levels of consciousness.

If you look closely at the essential elements our space has been built of, you will see the star tetrahedron, and an octahedron, and a tetrahedron, and, of course, individual spheres.

Evolutionarily, advanced space includes all previous shapes. Once we shift the attention (center) from the octahedron to the tetrahedron or any single particular sphere, the shape of space around the chosen center will change immediately. It's the same as shifting the point of view of the world from integrity to something separate that distorts an individual's vision of the world, including one's perception of reality. That is what intellect does; it generates an individual's ego. False space full of conventions is built around this center, although it looks quite real and corresponds to general laws. That is why humanity is being easily misled. Once we abandon intellect, our perception returns to a single center again, following natural laws, where nothing is separate. When there is no specific unit in the center, this place is occupied by nature itself because it is always present everywhere anyway. You've just seen it for yourself. When the ego leaves the center, then common integrity is manifested. Only together can we build a harmonious space following the natural laws of evolution, where we become gods ourselves. This is perhaps our most important lesson from this chapter.

## Chapter 4

# The Music of the Divine Spheres

*The song is full of lightsome weep,*
*So it's unclear in agonizing calm*
*Whether the soul has turned its voice to wind,*
*Or it's the wind that turned into the soul.*
José Santos Chocano

When we talk about harmony, we always mean that there is someone who evaluates this harmony because, from the point of view of physical nature, all ratios, proportions, and vibrations are equal and have the equal right to exist. So, harmony depends on the properties and features of the ones who draw conclusions about harmony, or rather on one's senses, as harmony is eventually being evaluated by the observer's senses. It is directly related to what conceives one's senses and feelings, to the substance we call the soul. The mind has an immediate influence on what the soul feels. For this reason, the peculiarities of perceiving harmony are directly related to the properties of space, especially space of the spirit and space of the mind. Just as our perception of harmony depends on the laws underlying these spaces' structure, the perception of other animate beings also depends on these laws. Scientists have proved that even plants grow and evolve better while listening to the sounds of harmonious music. That is, the harmonious external world contributes to the manifestation and evolution of life in it. Conversely, a disharmonious world can suppress and destroy life.

When you truly understand how the Tree of Life grows, then you can also understand the universality of the laws underlying this process, because these laws were developed by nature itself.

These laws operate and manifest themselves wherever there are space and vibrations. They're actually everywhere. They can be applied to light; they can be applied to sound. This thought came to me soon after the Tree of Life had revealed its secrets to me.

As it came, it completely changed my life for at least a few months. After all, everything is so simple: if these laws work with sound, then there are ratios between tones that make all of them sound in complete and perfect harmony with our consciousness. These ratios correspond to the ratios of the Divine Spheres! If the energy centers of spheres are located in space at an absolutely precise and certain distance from each other, then there are vibrations of certain frequencies between them. The frequencies of these vibrations should be proportional to these distances.

To understand the importance of this discovery, we need to recall the historical facts that have led to the evolution of music as we know it now. Let's first remember the history of music — that is, the history of patterns and ratios between tones and sounds that humanity has discovered and adopted, which has resulted in music sounding the way it does now.

It should be said that, at different times, different peoples had not only different ratios between tones, half-tones, and intervals but also a different number of basic notes: from five to several dozen. However, "more" doesn't mean "better." To understand this, let's turn to ancient Greece and see the works that Pythagoras and his disciples dedicated to music. They have actually laid the foundations of modern music.

Pythagoras was extraordinarily serious about studying simple whole numbers. We might even say that he divinized them — that is, instilled them with divine meaning. In his school, there were no zeros, negatives, or "any kind of" irrational and other numbers. He also attached great importance to the distinction between odd and even numbers, considering the first to present the active, or male, principle, and the second to present the passive, derivative, or female principle. It must be said that humanity realized the difference between these two principles a long time ago. In the heritage of Pythagoras, there is a clearly traced gleam of knowledge that had come to humanity from much more ancient times and had an impact on our modern ideas. Let us recall, for example, the ancient name of God from Kabbalah, which consisted of four letters: the first and the third stood for Father and Son respectively, the second and the fourth denoted Mother and Daughter—two active principles and two passive ones. However, whole numbers are also conditional because they do not exist in nature by themselves. They exist only in the human intellect. Anyway, let's go back to Pythagoras.

Pythagoras's love for whole numbers has also left an imprint on his works, devoted to the study of music and harmony. Later on, his disciples and followers went on with these research studies. These works and experiments have allowed the Pythagoreans to create pentatonic and diatonic musical scales. They have discovered the basic ratios between sound tones, which were pleasant to one's ear. Naturally, as you may have guessed, these were ratios between simple whole numbers, such as 1:2:3:4.

Thus, an octave had a ratio of 2:1—that is, the first note of each subsequent octave had a frequency that was exactly twice as high as the previous one. In addition, the "good" sound contained tones with ratios of 3:2 (the fifth tone, or, as it was later called, the "perfect fifth," due to its peculiarity among other tones and half-tones) and 4:3 (the fourth tone, or the "perfect fourth," which is considered to be an inverse harmonica of the fifth tone).

Pythagoras was apparently quite pleased with the result because he believed he had discovered the basic laws of the universe, where the creator apparently expressed himself.

Aristotle wrote, "The Pythagoreans have discovered that the ratios of the musical scale were expressed in numbers, and numbers seemed to be the first things in the whole of nature. Elements of numbers were elements of all things, and the whole heaven to be a musical scale and numbers."

The musical proportions discovered by Pythagoras were admired by many people. I have even found a photo of a building constructed with the musical scale proportions. Its architect was inspired by music. I must note that buildings constructed within the golden ratio proportions look much more elegant and harmonious.

The Pythagoreans believed that distances between planets were subject to the very proportions they had obtained, which were the same as the strings producing harmonious sounds. They thought that the solar system consisted of ten spheres orbiting the fire in their center, and each sphere uttered its own sound. The closer it was to the center, the lower the sound it uttered (note this conclusion). These sounds create a wonderful harmony called the music of the spheres.

Everything that has been said is absolutely correct. Spheres deserve even much finer epithets. Pythagoras knew somehow that it has to be exactly this way. My study of this issue suggests that Pythagoras received fragments of knowledge from the ancient Egyptians. In those days, many Greek scientists traveled to Egypt to acquire knowledge.

If Pythagoras had learned it himself based on his experiments with stretched strings, his conclusion about the tones of spheres sounding would have been completely opposite. After all, shorter strings produce higher sounds. It was these fragments of knowledge that pushed Pythagoras to seek the knowledge that would determine his entire life path.

There is one small detail: the Pythagoreans did not discover the real proportions of the Divine Spheres. If they had found them, they would have realized that geometric ratios and theorems, including the famous Pythagoras theorem, originated from the laws of the spheres. What they got based on their experiments was only a small occasional case, those few ratios that were generated by sounds pleasant to the ear.

Much later, in the sixteenth century, the musical scale was improved, having turned into the scale we know now. This upgrade divided each octave into twelve identical intervals (it included seven main tones and five half-tones, with the twelfth tone corresponding to the next octave). Each subsequent note sounded about 6 percent higher than the previous one. The advantage of the equivalent (or chromatic) scale was that it was the same for all keys and could be easily moved up and down. Therefore, musical keys could be changed easily in a piece of music. Humanity had failed to solve this issue before. In addition, its tones and half-tones corresponded to the ratios of whole numbers obtained by the Pythagoreans to within 1 percent. None of the existing musical scales could boast about it. For this reason, the half-tone scale has been adopted in the West for more than two hundred years.

Although these scales are well known and even standardized, none of them are perfect. Ultimately, these scales are not perfect enough to be called "those developed by the creator himself." Thus, they generate some combinations of sounds that are in apparent dissonance—that is, they set the teeth on edge of clear disharmony. For example, combinations of C-C#, C-D, C-F#, C-Bb, and C-B are clearly not consonant. Because there's nothing you can do about it on this musical scale, musicians have just avoided these unpleasant combinations and played the music that sounded good.

This happened because the Pythagoreans were too carried away by whole numbers, unfairly ignoring other numbers. Therefore, they did not understand that the real ratios were proportions of square roots of whole numbers, not the numbers themselves. However, this is not surprising at all.

Thoth once told Drunvalo that it took the Egyptians ten thousand years to realize that the one was not really the one, but a square root of one. Unfortunately, he gave no further explanation. So, by reading these chapters, you have saved at least fifteen thousand years of your life (just kidding).

The idea that there is a combination of tones that are impeccable and completely harmonious changed the ordinary course of events in my life for at least a few months. My feeling told me that, because the center of each sphere composing space is located at a strictly defined distance from the centers of other spheres, there must be certain unique vibrations between the spheres corresponding to these distances. The set of these vibrations should form an absolutely harmonious melody (in terms of consciousness and spirit evaluating this harmony).

While trying to combine ancient knowledge and modern harmony, I saw that the condition of the 2:1 octave ratio is also fulfilled for the second layer, for the next pair of layers (the third plus the fourth layers), for three layers (the third, plus the fourth, plus the fifth layers), and so on. This is due to the fact that the centers of spheres of the lower line in Fig. 6 correspond to those whole numbers Pythagoras loved so much. If you combine the scale with the third and fourth layers, the main Pythagoras proportions will point exactly to the centers of the tangential spheres. The first tone will correspond to the square root of four (i.e., two), and the last one will correspond to the square root of sixteen (i.e., four). The "perfect fifth" is called perfect because it gets exactly in the center of the intermediate sphere with the number "three." The "perfect fourth" is equal to the ratio of distances to the centers of the fourth and the third spheres. So, the 1:2:3:4 proportions of Pythagoras are actually proportions of $\sqrt{1}:\sqrt{4}:\sqrt{9}:\sqrt{16}$.

There is actually no need to constrain the scale with a rigid proportion of 2:1, because this ratio is also a special case. After all, each layer is actually a complete and absolutely harmonious scale, with the only difference being the number of tones. Thus, the scale of the first layer has only one note, corresponding to √1. In the scale of the second layer, there are three notes, (√2, √3, and √4). In the scale of the third layer, there are five notes, and the scale of the fourth layer has seven notes, and so on.

It even sounds pretty. For example, "this piece of music was written in the scale of the fifth layer" or "in the scale of the third and fourth layers." If you combine the scales of the third and fourth layers, you will get 5 + 7 = 12 tones. This number is the same as in the common scale for musicians, with the only, but significant, difference in proportions between tones.

To make it easier to define the frequencies of the Divine Sequence, I've compiled Table 1, which you will find in the "Additional Information" section at the end of the book. This information will help those who want to make instruments set up in the new musical scale. To make it more convenient, I have also placed an identical table, namely Table 2, recalculated respectively for the special frequency of 432 Hz.

Then I saw that there was a certain pattern in the formation of minor and major chords. In modern Western music, major and minor triads are formed through the combinations of tones with the ratio of frequencies equivalent to the ratio of simple whole numbers, such as 6:5:4, Fig. 27. Since minor triads are not so accurately expressed by ratios of these numbers, they are formed by other similar ratios of numbers, which are closer to the harmonic sequence: 15:12:10.

If you look at the example of how major and minor triads are formed in the traditional musical scale, Fig. 27, you will see the difference between them. If we take into account how intervals between levels change within a space of spheres, then the law of building a major triad will correspond to the "way up," from the center to the outside, with the increase in level numbers. A minor triad is a "way down," with the decrease in the level numbers. The spirit (your soul) grasps the laws and ratios very subtly and unmistakably and lets you know about it through your feelings. Due to this difference, we (or rather, our consciousness) subjectively feel different emotional impacts of these different triads.

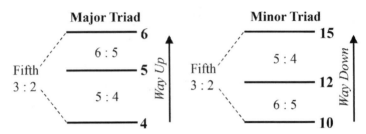

Fig. 27. The composition of intervals in traditional major and minor triads reminds the spirit of two possible paths of movement in the space of consciousness —ascending and descending. Therefore, it triggers two opposite feelings in our souls.

By its vibration, the minor triad reminds your soul of the way down, the path of decreasing the dimensionality, thus causing sadness, sorrow, yearning, and reverie. The major triad reminds us of the natural way up, the path of natural evolution. It creates joy, uplift, and excitement.

Table 1 forms the basis of another study I conducted: to find how the current chromatic scale differs from the harmonious scale of the space of spheres. To do this, I placed the tone frequencies of the standard chromatic scale in the figure (Fig. 28) along the vertical axis (with tones and half-tones corresponding to the piano keys), with a note A corresponding to the tone of 440 hertz. Then, I've tried to found a harmonious sequence of tones that is as close as possible to the standard musical scale. As you would expect, it turned out that the tones of the standard chromatic sequence are impossible to be included in one harmonious sequence of the space of spheres. However, it is possible to split the standard tones (approximately) between two harmonious sequences shown in the drawing as Sequence 1 and Sequence 2.

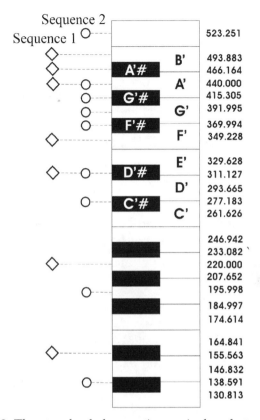

Fig. 28. The standard chromatic musical scale tones belong to two different harmonic sequences of spheres' space. Our hearing perceives the combination of tones belonging to different harmonic sequences as disharmonious.

Right away, we have found the reason why some combinations of tones of the standard musical scale sound bad, for only some tones of this scale exactly or roughly correspond to the same harmonious sequence (a sequence of square roots of whole numbers). Others belong to completely different sequences! I have marked tones belonging to the same harmonious sequence on the left in the figure. The tones marked with diamonds belong to the same sequence, and the tones marked with circles belong to the other one. This is the reason why some combinations of tones of the chromatic scale sound dissonant. Tones belonging to different harmonious sequences belong to completely different spaces of spheres, and therefore they sound disharmonious together. From this figure, you can easily understand why, for example, the above combinations of tones are perceived by the ear as disharmonious.

What we have just learned about harmonious vibrating sequences can be applied not only to music. These examples illustrate well what happens to humanity in real life. After all, each person initially has one's own pure harmonious sequence of vibrations, formed in exact accordance with the laws of life, the laws of the universe, and one's individual space of spheres with its own unique meter (or unit of measurement). This sequence is an integral attribute of your spirit or soul. Therefore, we can say that it was created by the creator.

However, in addition to this pure sequence, there are also other "tones and half-tones" belonging to vibrations of other people in the real world around us. They do not always belong to the Divine Sequence. The kind of vibrations depend on each individual, on their programs, and their affections. Some of them might have been passed on from one's ancestors. The stronger these programs and affections are, the louder these alien vibrations "sound."

A person (especially a careless person) can get so accustomed to them that they will not even distinguish which of them are inherent and which of them are not. It seems as though everything is unified, and the entire life sounds like a single continuous noise, which is difficult to understand. Often these external (coming from the environment in which a person lives) and internal (coming from the affections and programs of a person) vibrations are so strong that they drown out even the original sequence itself. If they conflict with the Divine Sequence's vibrations, they cause disharmony and dissonance, resulting in disharmony in a physical condition that directly affects human health.

In this regard, we can distinguish possible effective treatment methods, as they target precisely the cause of the disease and basically aim at weakening the induced external and internal destructive vibrations (by changing the external environment and transfiguring internal programs and erroneous beliefs), either by strengthening the initial vibrations or by switching the person's attention to them. Sometimes they might combine both methods at the same time. The expression "initial vibrations are strong" means only that the induced ones are weak in comparison with them, and they are only a minor background for this person, without having any significant impact. One can easily cope with their influence.

However, for the average person, this, unfortunately, is not so easy, as disharmony can be quite strong. In this case, external vibrations will be able to help: sound, light, and possibly other kinds (not visible and not tangible by our senses), selected and composed in accordance with the law of universal harmony. Eventually, the treatment result will depend on how much it will be possible to clear the original sequence from harmful vibrations. Physical health is directly related to spiritual health. This is not a hypothesis but a truth. This was well known to ancient Egyptians because they used sounds and colors to heal people. Fragments of this ancient knowledge are still being found today. For example, there are old methods and new special methods of treatment that utilize different combinations of colors. In fact, the holistic knowledge of this subject was completely lost by humanity.

It should be noted that the standard A tone (440 hertz) in modern music is far from being the best or perfect. Moreover, according to musician and researcher Brian Collins:

the modern music setting based on A = 440 Hz does not harmonize with any level corresponding to cosmic motion, rhythm, or natural vibration. Mozart and Verdi both based their music on the natural vibration of A = 432 Hz, which was called "Verdi's Settings." Most Western music, including new age music, is still tuned to the unnatural tone of A = 440 Hz. The difference between A = 440 Hz and A = 432 Hz is only eight vibrations per second, but this results in a significant difference in perception at the level of consciousness.

On the internet, you may find yourself and compare the sound of the same piece of music performed with A = 432 Hz and A = 440 Hz, and sense the difference in perception. The musical tone of 432 Hz is even used in musical therapy. Sacred songs and mantras are often performed with the setting on A = 432 Hz as a "divine tone."

According to Robert Lewis, a disciple from the Brothers of the Rosy Cross, "the purpose of music in religious service is to raise the vibrational level to the level of the spirit through a few overtones." According to some authors, archaic Egyptian instruments tended to have been set to A = 432 Hz. In ancient Greece, musical instruments were mostly tuned to A = 432 Hz. The Greeks believed that Orpheus, the god of music, was the keeper of Ambrosia and the music of transformation. His instruments were tuned to a frequency of 432 Hz.

Some authors suggest that it is possible to make sound harmonics such as 72 Hz (9 x 8 Hz), 144 Hz (18 x 8 Hz), and 432 Hz. Then, they can be synchronized as music in the binaural 8 Hz to "awake the orchestra of our thoughts in the cathedral of our mind." They claim that such musical harmonics can also resonate, leading exponentially to the harmonization of amino acid metabolism time within a double helix of DNA. Another harmonic alternative is called A = 444 Hz (C (5) = 528 Hz) because this frequency is more compatible with nature than the frequency of 440 Hz. Curiously, now we can figure out the reason why the frequency C (5) was chosen to be 528 Hz. It's simple: C (5) = A $\sqrt{2}$.

Religious leaders have simply replaced the original musical scale containing A = 444 Hz and C (5) = 528 Hz, and a wonderful "E," the tone played by the Pythagoreans, with the current scale using A = 440 Hz. This knowledge was destroyed for centuries.

Leonard Horowitz suggested that instruments and voices tuned to the frequency A = 444 Hz are much more acoustically pleasant, instinctively attractive, mentally refreshing, and scientifically associated with genetic recovery [3]. They can even carry the vibrations of pure love. Frequency A = 440 Hz has historically been associated with monopolization of the music industry and the military commercialization of music. Due to the Rockefeller Foundation and the US Navy, the 440 Hz frequency was introduced as a standard into military orchestras and then into all other kinds of music.

L. Horowitz believed that this was done intentionally because this frequency is able to cause and maintain a sense of aggression among people subconsciously; it can be used for psychosocial agitation, as well as cause psychological stress, leading to various diseases and, therefore, to the increase in profits of monopolies. Therefore, the biologically natural frequency of 444 Hz was replaced by 440 Hz due to censorship. Thus, to promote health and peace, a musical revolution is needed. This revolution has already begun with musicians who are bringing instruments back to the optimal sound to ensure positive effects on their audiences and bring back the unity of art and science.

I have managed to discover that the entire modern standard musical scale is built according to the law of "depressive" musical sequences, dismally affecting human consciousness and mental health. It means that although all existing Western music, from sophisticated classical music to military marches, can trigger different feelings and experiences according to the composer's initial ideas, this music also has a deep-level and oppressive influence over the human spirit.

You'll understand it easily if you look at Fig. 29. This figure compares the laws of composing octaves within the standard musical sequence (bottom curve) and the sequence of Divine Spheres (upper curve). Each octave has exactly 12 tones (the thirteenth tone corresponds to the first tone of the next octave). The 12 tones of the Divine Sequence correspond to the second (five tones) and the third (seven tones) layers combined together. Both curves show how many times the frequency of each tone in the corresponding octave is higher than the frequency of the first tone of the octave. The figure shows that, within the existing musical scale, the distance (in frequency) between tones increases with the increase in the tone number, which corresponds to the "way down." In the Divine Sequence, contrariwise, the distances between tones are reduced with the increase in the tone number, which corresponds to the natural "way up." As we already know, spirit unmistakably senses intervals and responds accordingly. I suspect that the practice of composing "depressive" musical scales dates back to the late period of ancient Egypt, when the remnants of knowledge still existed but were already being used against humans to control (or rather, to lower) the consciousness of the populace. It is clear why it was necessary: a person with a suppressed spirit was easier to manipulate. Religion serves the same purpose.

A lot of people will probably say, "Are you serious?! Why, then, are there so many beautiful works of music that indulge our soul? After all, they are written in the so-called "depressive scale." It's a misconception!" In fact, there is no misconception. While experimenting with music, I learned to distinguish two different things that others don't usually notice. A piece of music has two effects on our consciousness. One effect is produced by the skill, feelings, and emotions of the composer, which they have nested in a piece of music. The analogy is the feelings and skills of an artist, embedded in the canvas. The second effect is produced by the combination of tones itself—that is, by the musical octave. The analogy is a color palette chosen by an artist to paint the canvas. When the first effect is strong, humanity does not notice the second one. However, as soon as you listen to just random combinations of tones, the effect will be quite different. Imagine how strong a work written by a talented author in the scale resonating with consciousness could be!

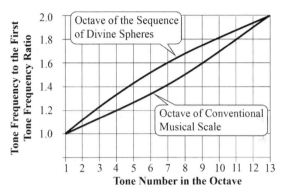

Fig. 29. A comparison of the composing laws of the standard tempered musical scale and the sequence of Divine Spheres shows that the conventional musical scale is built according to the laws of "depressive" musical sequences oppressing the human spirit.

The Laws of the Divine Spheres operate not only with the sound. If you take light waves with frequencies corresponding to the same ratios of square roots of whole numbers, you'll get a palette of an impeccable octave (or scale) of colors, on the basis of which you can paint pictures delighting your spirit. The soul of every person will enjoy this scale because it is chosen according to the laws your soul lives within. Even if you mix these colors with each other in any combination and proportion, the scale will still remain perfect. The only thing you need is to choose or create chemicals or minerals that reflect or transmit light waves of certain frequencies, and then choose the substances or minerals that reflect or transmit light, with frequencies belonging to the same sequence.

However, that's not all. By using these laws, you can paint a picture in minor tones, or in major tones, depending on what feelings you want to convey to the one who will look at your canvas. It's the same as in music. You'll be able to paint pictures or make stained-glass windows whose contemplation will elevate your soul as well as the souls of other people closer to heaven—that is, to the highest levels of consciousness. You can do everything in color.

You can also complement music with light (permanent or changing in color and/or intensity) made up of the colors belonging to the same octave of the Divine Sequence, and your soul will thank you for this wonderful, unforgettable gift as it is set free from the weight of everyday life and reminded where its real home is. You can do real miracles with all this.

The color scheme can be selected in the same way as the sound scheme. The only difference is that the frequency of sound is measured in hertz, while the frequency of visible light is measured in terahertz (THz). Each harmonic frequency series, expressed in THz, has its own sequence of color tones. It's just physics. For example, the frequency of 440 THz corresponds to a specific tone of red color. The frequency of 580 THz is a specific tone of green color; and the frequency of 622 THz is a specific tone of blue color. Verification of this law in experiments showed that the selected color series and minor and major color triads evoke feelings in a person similar to the feelings that one experiences when listening to the sound corresponding to them. Unfortunately, in this book, I cannot give you the colors and color gammas, but you can find examples on the milovanovbooks.com website.

It should be said at once that computer screens and printers, both laser-beam and ink-jet, are unable to convey actual colors. The fact is that a computer screen and a printer only create the illusion of color. They create this illusion through different combinations of a limited set of certain colors, often three: red, green, and blue. That's why the pictures in a computer screen or in a printed book convey colors only approximately. Actual colors can only be obtained by the light source with a single wavelength, such as lasers, or by using substances that reflect or transmit certain wavelengths only. Then actual pure colors will look something like they look in a rainbow or in a rainbow obtained by a prism that refracts sunlight. Paintings made in actual colors and their combinations will be perceived in a completely different way, not like ordinary paintings.

I was able to find some examples of stained-glass windows whose authors have intuitively picked up colors whose combination makes an unusually deep impression on a person. The reason for this effect is that the stained-glass windows were made in a scale of colors close to the scale of the Divine Spheres. Imagine the impression a painting or a stained-glass window can make on a person's consciousness, if the colors are chosen exactly in accordance with the scale of square roots of whole numbers.

One such example are the large stained-glass windows in the Sagrada Familia temple of Barcelona, which were made in a color scheme close to the sequence of Divine Spheres. It's like it takes the viewer's consciousness to other, higher, dimensions. If these beautiful stained-glass windows had pictures showing the real laws of the universe instead of religious plots, such a temple would allow us to experience truly amazing feelings. Unfortunately, these natural laws can be used not only for good. These stained-glass windows are at the same time an example of how natural mechanisms are skillfully used by manipulators of human consciousness, those whose purpose is to implant ideas in human minds (in this case, it is religious and ideological), which induce people to do what is beneficial to manipulators. The trick is to use the following natural mechanism.

When people hear or see something that resonates with their consciousness and the laws of the universe, which have been imprinted in human genetics over the course of evolution, they feel a release, and their natural inner protective mechanism of consciousness weakens at this moment. This may happen, for example, when they come into contact with nature or its laws, see a beautiful painting, hear the singing of a choir, or the sound of bells, or an organ, and so on. This is a natural process. The danger lies in the fact that the "viral" ideological program can be pretty easily induced in the human consciousness in such a disclosed state, destroying, blocking, or changing the consciousness of a person in a way beneficial to manipulators.

For example, this is how churchmen use this mechanism. They first speak to you, listing a set of rules our ancestors knew and honored back in those days when there were no religions on Earth at all, but their worldview reflected the laws of the universe. Churchmen list truths such as "don't kill," "don't steal," and so on, which have been present in human consciousness at the genetic level for hundreds of thousands or millions of years. It is clear that people, unless they have their own much more evolved analytical apparatus, are imbued with trust, naively believing that the speaker of such correct things cannot deceive. This is wrong. It is possible for people to lie, and church manipulators immediately and imperceptibly complete this list of capital truths with absolutely alien, unnatural ideas that destroy consciousness and turn people into obedient slaves. For example, they tell you:

> You are a slave (of God), a lamb (i.e., a will-less sheep whose owner master can use however he likes). Everything that happens is the will of God, so you must accept it humbly and resignedly. You must not fight or stand up to your enemies; you must "love" them and tolerate everything they do, turn the other cheek, take a hands-off approach...

...and so on.

Moreover, the churchmen and their masters are very persistent in imposing such ideas able to weaken countries and nations, to nestle them in human consciousness. For example, in the territory of Kievan Rus, three-quarters of its population were killed to force people to accept Christianity. Such criminal introduction of Christianity is now interpreted as the merit of Prince Vladimir. Everything has been turned upside down in our parasitic world. Betrayal and crime against people are interpreted as feat and merit, and protection of the fatherland is declared crime and robbery. Here is another typical example: Alexander the Great and his many troops sidestepped Armenia because he was afraid of engaging in battle with Armenian soldiers, who were famous for their strength, skill, and endurance. After the adoption of Christianity, only one-twelfth of the original territory was left in Armenia, and the remnants of the people were subjected to genocide in no time. The same thing has happened and is still happening with other nations, including Russia. As another example, churchmen may also instill in you the idea that you are exceptional or "faithful" and that everyone else is either "pagan" or "second grade," and therefore, they can be killed or used as slaves with impunity.

The brain is arranged in a way that, when it perceives a list where there are lots of good and correct things as a good list, it often "swallows" the entire list without analysis, believing that all the components of it are good as well. This is the devilry of ideological poison.

The effect on consciousness is multiplied by the "comprehensive approach" when, during such instillation, a person is additionally affected by sound (e.g., choir singing, sound of an organ or bells), light (for example, using candles or lamps), colors and proportions (as in the case of the stained-glass windows mentioned above), or using the effect of the crowd. Manipulators use similar methods, as well as many others. They do about the same thing to your consciousness, as the so-called "genetic engineers" do to natural genes, turning healthy biological organisms into unviable ones, which has destructive effects on humanity and life on the planet in general.

If you carelessly swallow their "modified" ideological product, then about the same thing happens to you as when you swallow a genetically modified food product: it feels nice; it quenches hunger (in the first case it's physical, and in the second case, it's informational), but, at the same time, it quickly brings death, not only yours but also your nation's. These examples show that such "genetic modifications" of consciousness are in fact not a harmless prank but a powerful ideological weapon, much more dangerous and destructive than nuclear weapons because they are capable of weakening and destroying entire countries and nations.

A similar mechanism is used by ideologists, politicians, and even advertisers—by all who want to manipulate your consciousness for profit, triggering you to commit actions and moves you would never commit in a normal state. Unfortunately, most people do not notice this mechanism of criminal influence on their consciousness. However, its consequences are devastating because they cause the degradation of human consciousness and society.

Therefore, to not fall into this trap, every time you experience a deep feeling inspired by an activity (which can be manifested in words, sound, color, and proportions), it is important to pay attention also to the content that is nested in this activity by those who offer it to you.

Overall, I hope that this small digression will allow you to avoid serious mistakes in choosing a life position and life path. Let's continue our conversation about the Laws of Spheres. When people learn to identify the vibrational frequencies of different smells, they will be able to easily make their harmonious combinations and aromas on the basis of this law. These laws are so all-embracing that, if you ever see the light coming from spiritual beings, I'm sure it will be based on these same ratios.

When I realized this, the greatness of the Laws of Spheres flooded upon me with a wave of experiences that could not be described on paper.

A few days later, when the excitement in my soul subsided a little, I set out to test my findings in practice. It was not an idea, not an assumption, and not a theory—it was a certainty based on knowledge, sense, and understanding.

The knowledge of this had already been present within me; the test was triggered by my childish curiosity rather than necessity. I was extremely excited to hear the sound of the Divine Spheres.

How was it better to implement it? Because I did not have a sound generator, I started surfing the internet for computer software that would replace it. Finally, I downloaded one of the most attention-focused software programs to my computer. I didn't know yet how this program worked, but it allowed setting and changing one tone modulated in some complex way, as judged by the given mathematical formula. It was enough to start with. I launched the software and typed a number corresponding to the frequency of the first tone I had calculated. I had to press the key to hear the sound, but for some reason, I was taking my time. This moment meant so much, not only to me but also to all humankind as well. It was like waiting to hear the voice of the Superior Mind.

You'll easily understand my extraordinary excitement at that moment. After all, I was the first person in several thousand years, and maybe the first in the history of human civilization, who would hear the tones sounding in harmony with the Divine Spheres. You'll understand how I felt even better when you find out that the first sounds from the Divine Sequence were the sounds of bells. It's hard for me to describe the feelings I experienced; you have to go through it by yourself. The bells sounded within the perfect scale. It was just gorgeous!

To appreciate the importance of this moment, you must understand that nothing happens accidentally in our lives, especially in such moments. The arrival of the ancient miraculous icon at my house shortly before is strong proof. The form chosen by the sound harmony of the Divine Spheres to reach out to the human ear for the first time was not a coincidence. Those who have passed on this knowledge to us make it clear in every detail that it is sacred and secret. Please appreciate this amazing gift from heaven, which should be touched only by a pure heart.

A couple of days after this first experiment, which was conducted in April 2006, I was able to find the software that allowed generating combinations of several various tones within the new sequence. It allowed combining up to sixteen tones together and even listening to them in stereo sound! That was exactly what was needed.

According to the rules, I probably should have checked the sound of tones, starting from the first one, moving on to the second, and so on. But I couldn't resist. I set seven notes at the same time as I had calculated them in the ratios of seven spheres of the fourth layer, or Seven Divine Spheres. What I heard is indescribable in words. I felt as if hundreds of angels were calling me at the same time. I froze at my computer and probably listened to this sound for a few minutes because I felt as if it was addressing me. I couldn't shake it off and overcome the feeling of hearing something familiar but forgotten a long time ago. This feeling was triggered by a combination of seven simple tones! You can imagine what one could feel listening to a good musical instrument. I thought later that if a chorus had sung within this scale, you would have probably risen above the ground after your soul, rising into the sky. It seemed to me that my soul heard something it longed to hear long ago. It is impossible either to forget or to describe in words the feelings I have experienced.

Then I started checking the mix of different tones and their different combinations, and I didn't find a single unpleasant sound. All the tones perfectly combined with each other and sounded pleasant to the ear.

In the end, I checked the simultaneous sounding of thirty-two tones at once — the largest possible amount that could be played using the software. This experiment led to the discovery of a slightly different kind. I got an answer to a question I had when I was a kid. We are all familiar with the expression "ringing silence." No one can explain why, when you are in absolute silence, you hear not silence, but incomprehensible ringing in your ears. Well, that is how your spheres sound!

## Chapter 5

# An Experiment with the Music of the Divine Spheres

Some readers might wonder whether there might be any practical personal benefit of understanding the Laws of Spheres in their daily lives. In fact, there are, and, in this chapter, I will tell you what benefits you can get. It was for the purpose of answering this question that I conducted my research. To better understand this, let's first look at the environment the vast majority of people live in and how it affects their consciousness.

Ancient sources say, and my research confirms it, that each person has several levels of consciousness. There are higher and lower levels. Under the influence of desires and aspirations, one's consciousness can move to certain levels and activate them. That is, either higher or lower levels can be activated, with corresponding consequences for the person and the world around them. Many people understand this and sense it intuitively. Those who have the ability to feel the energy flows in their body can physically feel this process as the activation of a particular energy chakra. For example, when a person experiences selfishness, fear, anger, greed, depression, dismay, and other negative feelings or is a slave to their desires, their consciousness automatically "falls" to lower, semi-animal levels, where the person lives at the level of instinct. One's actions are predictable and easy to manage. Conversely, when a person is driven by the desire to help others, to create, to make the world better, then one's consciousness moves to higher levels—the factor that makes people different from animals.

While being at these levels, people are able to create, make more reasonable decisions, experience superior human feelings, better understand the view of the world and difficult reality, and distinguish what is truly valuable and what is false. Such people are more conscious — that is, they have a higher level of consciousness, which makes them independent and difficult to manipulate.

Different people may have different amount of consciousness levels, depending on the evolution level of their spirit. Thus, the primary purpose of our stay on Earth is evolution of our consciousness and the development of its new levels. All that a person's spirit will leave this world with is one's either evolved or degenerated consciousness, depending on how the person lived. Consciousness is arranged in such a way that it and its new levels are evolved only when a person is active on one's highest levels. It is a natural process for normal civilizations. If a person is at lower levels of consciousness most of their life, then their higher levels (developed throughout their past lives' evolution) will gradually collapse. In this case, people did not fulfill their primary goal, defined by the laws of nature. They went against these laws, destroying themselves. This is important to understand.

Unfortunately, in the world we have on Earth now, many people live on lower levels of consciousness, missing out on huge opportunities and causing irreparable harm to themselves and their consciousness. Let's look at our reality. A small group of people parasitizing someone else's work and potential, the so-called elites who seized power once, have taken thousands of years to come up with new methods of reducing human consciousness to the lowest levels and keeping it there. Why are they doing this? Because they understand they can hold their power only by having control over human consciousness. It is possible to manipulate only people who are at lower levels of consciousness. When people are frightened, depressed, and uncertain about the future, when they are economically on the verge of survival, when they do not have time to think about superior laws, they are easily manipulated. In such a condition, with the help of a stick and carrot, they can be easily forced to make decisions that are beneficial for their rulers. Institutes of psychology, which work for governments, consider that if you pour a certain dose of negativity on a person, it leads the person to disconnect from social life and cease to pose a threat to the system.

In this state, human consciousness falls to its lowest level, the level of instincts and primitive behavior. The methods used to degenerate our consciousness are easily traceable. It is also a system of constant negative news; movies that influence our consciousness negatively; an imposed system of false representations about the world and us in this world, especially false history; distorted education and science; a false system of values in society; imposed primitive music and the idea of sex replacing the natural feeling of love; violent sports; and many other methods, including destruction of consciousness by alcohol, tobacco, drugs, horror movies, rock music, and psychotronic generators.

As a result of the long-term use of these methods, humanity has been immersed in artificial, toxic relationships, a poisoned informational environment, and a life to which they are trying to adapt in different ways. In this environment, even a conscious and understanding person can sometimes, unnoticeable to themselves, fall to levels of consciousness they should avoid, where they become easily vulnerable and can make many erroneous decisions. Why is it harmful to be at lower levels of consciousness? Because when you're at lower levels, you do not evolve. In contrast, you destroy your existing higher levels of consciousness. You degenerate evolutionarily, destroying the accumulated evolutionary potential. You are not fulfilling the goal you have come to this world with. All this not only leaves modern rulers indifferent; it suits them quite well. Because this is their main goal. However, this should bother everyone who cares about their own consciousness and their own life. In these conditions, the responsibility for the right evolution of your consciousness and the consciousness of your children and for maintaining a positive attitude and creative state of mind lies only with you.

Ancient civilizations knew about the danger of staying at lower levels of consciousness for too long, and they also knew the methods and were able to make instruments to activate its higher levels. In the book, you will find all the necessary information and description of how to make simple but effective sound instruments that help not only activate the existing higher levels of consciousness but also stimulate your spirit to evolve new levels of consciousness. Visual tools can also be applied according to the same law for this purpose; however, technically they are somewhat more complicated.

Having conducted a lot of experiments with the Music of the Divine Spheres, to my surprise, I noticed that these sounds had improved my perception of the world and music in general in many ways. You could say that they have improved me in certain ways.

Having noticed this effect, in 2015, I decided to conduct a major experiment to test what effect this musical sequence has on people when resonating with consciousness. For this purpose, I made three simple, but, as it turned out, very effective instruments based on a well-known musical instrument called the wind chime. This instrument consists of a few metal tubes suspended on threads, tuned to certain tones. Why have I chosen this particular instrument? First, its sound is natural and very similar to the sound of a bell. A bell is an instrument with a powerful sound. Humanity is used to associating bells with the church. However, bells don't belong to the church. The church only adopted this instrument (as well as many other things) to increase its control on human consciousness. Bells don't even belong to this civilization; they have existed on Earth for hundreds of thousands of years. Although the church appropriated bells, the knowledge of how to use them correctly was completely lost or deliberately destroyed. This book restores this lost knowledge for the first time in 5000 years.

Second, I have set the task to make an inexpensive and compact sound instrument that each person could buy or manufacture and put even in a small room. The wind chime instrument meets all these requirements. In addition, you can hang it outside and allow the wind to play its music for you or put it inside and play it yourself. The latter method has turned out to be the most effective. Three different instruments used in this experiment are shown in Fig. 30.

Fig. 30. Three sound instruments: the "Voice of Consciousness" with a base frequency of 432 Hz (left) and a second tone of 432 Hz (center), and "the Call of Angels" (right) designed for a sequence with a base frequency of 432 Hz and tested on dozens of volunteers.

I have called the first two instruments "the Voice of Consciousness." They were tuned to the sound tones, which were in proportions of the first three layers of the space of spheres — that is, in proportions of the levels of consciousness a person had or close to these levels. The purpose of this type of instrument was to activate the existing higher levels of consciousness of a person listening to it. When such an instrument sounded, an unusual spatial "cosmic" sound appeared, independent of the individual tones sounding. A unique spatial "voice" of the instrument appeared, and that was why I gave them such a name. The first of these two instruments is tuned to the first eight tones of the Divine Sequence, with a base frequency of 432 Hz, and the second one is tuned to the tones from the second to ninth, starting from the tone of 432 Hz. The effect they produced was roughly the same. I already explained my choice of the original frequency of 432 Hz before. The tone frequencies of both instruments can be found in the first and second rows of Table 2 at the end of the book, respectively.

I have called the third instrument "the Call of Angels." It was set up in the proportion of the fourth layer of consciousness (the proportion of the Djoser's pyramid or the Seven Divine Spheres) — that is, those levels of consciousness that almost all people on this planet do not have yet. This instrument continued the tone sequence of the first instrument (with a base frequency of 432 Hz), and therefore it perfectly combined and complemented it. The sound of this instrument was unmatched. When it started, an indescribable "cosmic" sound appeared, penetrating every cell of your body. Space itself sounded like angels singing. That is why I gave it such a name. The purpose of this instrument was to stimulate the spirit of a person to work out new levels of consciousness that people do not have yet. Because it is possible to develop new levels of consciousness only in the process of vibrant creative activity aimed at changing the world for the better and in no other way, this device automatically stimulates people to do an activity. Therefore, it will be especially useful for people who are in a state of depression, decay, or just laziness. This instrument is ultimately quite powerful.

Unfortunately, conventional recording devices aren't able to convey all the sound effects produced due to their limited frequency and amplitude range. For example, when all eight instruments' pipes sound, they sometimes create peaks of an acoustic signal that are eight times the size of a single pipe signal. Conventional recording equipment simply cuts off such a signal. As a result, you can hear only something resembling the sound of an instrument. The "live" instrument produces a very different effect.

Next, my friends and I tested these instruments on volunteers in Los Angeles, where I lived at the time, and where the audience was very skeptical, as is probable in all megacities. But we must pay tribute to people who have willingly agreed to participate in the experiment. In total, several dozen people took part in it. To maintain the experimental integrity and exclude the factor of instillation, we did not tell the participants about either levels of consciousness or the new law. We just asked the volunteers to play a musical instrument and to describe in writing the changes or feelings they experienced in themselves. I was well aware that all people were different, with different levels and abilities of perception and ability to convey feelings. Thus, I expected that some percentage of people would not experience anything unusual. However, the experiment result exceeded all expectations. Although different people described their feelings differently, which was interesting and natural itself, the experiment had a positive effect in 100 percent of cases. Here are some of the most typical feedback examples from the experiment participants about the first instrument.

I like this sound. It doesn't take away my energy; it resonates within me. I could live with this sound; it is gentle, but not excessive.
Nancy N

I'm describing my immediate reaction to the black wind chime instrument: the energy came from the area of my shoulder blade, rose up along my spine, and discharged into my crown chakra.
Ley S., reiki master

The wind chime instrument is amazing. The sound is focused and seems to give a breath of fresh air. It soothes the mind and brings calmness and solutions.
Wane P, musician

My first reaction to hearing the Wind Chime was joy. It brought a smile to my face.
Katy K

This instrument has surprisingly clear, transparent tones. I've never heard such tones. There is something in them that is just naturally calming. On the day I heard them, I was under real stress. Always ready to put aside current tasks, I hung it in my office and started experimenting with this instrument, hitting the pipes randomly, with different intensity and frequency. As soon as I started playing it, I felt my stress just fading away. Oh my! It has a great sound, and, at the same time, it produces a therapeutic effect—what a pleasure!
John H

It was much easier with the third instrument. Impressions of almost all people after a few seconds of listening were similar to the following typical review:

Wow! I feel the energy of this sound, but I don't have the vocabulary to express it.
John H

This reaction is understandable. After all, not only do people not have the necessary vocabulary to describe sensations of this sound, but their spirit doesn't have it as well because although it feels the power of the Laws of Spheres, it has nothing to compare it with: it has yet to evolve the levels of consciousness in these proportions in the future.

The latter instrument is associated with a curious case. One doctor had heard about my extraordinary instruments and was eager to test them for himself. So he asked me if he could borrow them for a while. At that time, I had only this third, stimulating instrument. I lent it to him without a detailed explanation, believing that the doctor just wanted to listen to its sound in a calm home environment. I didn't expect him to be a diligent and responsible tester. When I saw him a few days later, he just ran up to me in excitement and relayed an eager tale of how he had hung the instrument outside in a strong wind and was sleepless all night because of this sound. The instrument affected not only him but all his neighbors around. I apologized to the doctor for not giving him detailed instructions on how to use the instrument and explained that the "Voice of Consciousness" could be listened to as much as one liked; however, one should be careful with the "Call of Angels." I usually recommend starting to listen to it for a few seconds a day, observing the effect, and then gradually increasing the time to a minute or more. It should also be taken into account that some effects of the instruments' impact can be felt immediately, and some appear after a while, after 1 or 2 weeks or even later. This is because time flows differently at the spiritual level. When the doctor found this out, he thought about it and asked my permission to keep the instrument for a few more days. During our next meeting, he said that the instrument really had the effect of a strong coffee. After testing both instruments, the doctor himself made me the following written recommendation on the official form:

I've had the pleasure of listening to the wind chime instruments developed by Alexander Milovanov. I find these instruments very pleasant and comfortable to hear. After a few minutes, I notice a feeling of calm and well-being. I strongly recommend Mr Milovanov's wind chimes for their soothing sound and mood-enhancing qualities.
David Starr (DC, QME, FICC)

When I started analyzing why the effect of both instruments was so strong, I realized that the impact of this sound on consciousness occurred directly at the spiritual level, bypassing programs of the intellect. Therefore, the positive effect was felt by all people, regardless of their upbringing, education, ability to perceive, and so on. The experiment involved mostly random people, except for one person. I intentionally asked Drunvalo Melchizedek to evaluate the instrument because he has greater abilities than ordinary people. If you will read more about this person on the internet, you will better understand what I mean. When Drunvalo listened to this instrument, he was simply delighted with its effect and especially interested in the mathematics underlying the sound it produced. After all, it was exactly the mathematics he'd been looking for.

We can say that these instruments work like an alarm clock for the human soul, or rather, the spirit. When the spirit hears the sound produced by the instruments, it instantly recognizes what's going on. Our spirit, unlike our intellect, instantly evaluates proportions between sounds, objects, and colors, bypassing all processes of the intellect. When hearing the tones resonating with the structure of its consciousness, the spirit instantly remembers what it is and what its main goals are. It immediately wakes up from the hibernation into which people had plunged it through the intoxication of habitual life: the routine of work, fears, anxiety, abnormal relationships between people, and so on. Having woken up, the spirit is immediately activated. It happens within seconds. Therefore, even a few seconds of playing were enough to obtain the described effect.

The knowledge and instruments described in this book will help humanity better withstand mind-destroying dangers, quickly emerge from stress and depression and maintain a positive and active state of mind. If this book and instruments allow people to raise their level of human consciousness at least a little bit, then my goal will be achieved. Sure, this knowledge and these instruments won't be able to solve all the world's problems. For example, if a person's intellectual programs are extremely strong, then no "Voice of Consciousness" will help. We need more radical methods to return people to normal. However, I'm sure it could help a lot of people. After all, it is the consciousness of people, not politicians, the government, or anyone else that determines how people live. In any case, it is not bad at all to have an alarm clock for your soul at hand.

## Chapter 6

# The Latest Scientific Evidence Confirming the Laws of Spheres

This might be a legitimate question: are there any modern scientific facts that directly or indirectly confirm the Laws of Spheres, except ancient artifacts and mathematics? It turns out that there are such facts. Let's turn to the latest scientific sources. The only scientist one can fully trust in this matter is the academician Nikolai Levashov. Why him only?

There are several good reasons. First, he has extensive scientific knowledge in many fields, and he obtained and tested all the knowledge himself without relying on outside influences. Second, he possesses abilities, such as traveling in space and time and control of nature's forces, which other people on our planet could only dream of. Third, he never lies. Due to his abilities, he obtained and gave humanity knowledge that was millennia ahead of science. Unfortunately, the value of this knowledge is still underestimated by contemporaries, as often happens with knowledge that is ahead of its time. However, our descendants will undoubtedly pay tribute to this scientist's genius and what he has done for humanity. In the meantime, unfortunately, humankind does not fully appreciate the fact that they are only alive on this planet thanks to Nikolai Levashov.

Human beings are arranged in a strange way. For thousands of years, they dreamed of unraveling the mysteries of life and the universe. Here comes the Star Genius, Levashov, who reveals these mysteries, as well as the mysteries of what is happening outside our universe. He reveals also what is happening on our planet, but many people do not even want to read his books. Let's leave it to the conscience of people who live according to their levels of consciousness. I will only say that if you

117

have not yet had the opportunity to read the books of Nikolai Levashov then try not to miss the available unique chance in this incarnation on Earth to learn information that will advance the evolution of your spirit thousands of years forward.

Let's look at how levels of consciousness are related to the knowledge provided by Nikolai Levashov. The structure of levels of human consciousness is determined by the Laws of Spheres. Each of these levels does not exist on its own. Each of them has its own material medium. Material media of consciousness are bodies of the human spirit. According to Nikolai Levashov, each person has several other nested bodies of spirit besides their physical body. They are composed of various combinations of seven primary forms of matter. Levashov marked seven primary forms of matter, which compose not only bodies of spirit but also our entire universe, with Latin letters A, B,...G. Various people can have various numbers of spiritual bodies, depending on their level of evolution and, accordingly, their level of consciousness. The order of spiritual bodies is the following: a physical body is followed by one ethereal, two astral, and four mental bodies. According to Nikolai Levashov, when a person evolves all these bodies, including four mental ones, over the course of one's evolution and gains experience, then "all qualitative barriers of the planet disappear for such a person. The Earth's zero cycle of spiritual evolution ends, and the stage of cosmic evolution begins. The presence of mental bodies gives the person possessing them tremendous mental power whereby they can influence processes taking place in nature, both locally and on a planetary scale. Through the sole power of thought, they can influence and control the processes taking place in human society; to see and to hear the past, present, and future; to influence and change the future of both separate individuals and humanity in general; and to move in space and the universe. Such people are capable of much, much more. Only people who are on the right path of evolution, the path of goodness, can reach this level. The evil, no matter how strong it may seem at first glance, is incapable of evolution" [4]. Such a person also has the opportunity to choose any planet in our universe for life and further evolution. Sure, there are only a few such people on our planet. But at least every adult has a physical, ethereal, and one astral body. In Fig. 31, I have placed a sequence of bodies of the human spirit, as depicted in Nikolai Levashov's book *The Last Appeal to Humanity*. I've only slightly changed the order of numbering these bodies, placing the ethereal body behind the physical one, so that all eight of them can be visible.

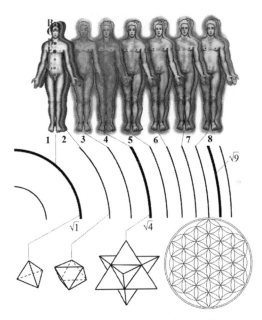

Fig. 31. The relationship between bodies of human spirit, the levels of consciousness, and geometric shapes corresponding to these levels.

This sequence includes the following bodies: physical (1), ethereal (2), first astral (3), second astral (4), and mental bodies from the first to the fourth (5–8), respectively. A diagram of human consciousness levels corresponding to the Laws of Spheres is placed under the bodies of spirit. Geometric shapes corresponding to these levels of consciousness are placed under the diagram. In this drawing, it is clearly visible why a person has exactly one ethereal, exactly two astral, and exactly four mental bodies. Each type of body belongs to a specific layer of consciousness, consisting of several levels. Each new developed body of spirit adds a new level of consciousness. It should be noted that both bodies of spirit and levels of consciousness cannot exist separately but are embedded in each other and represent integrity. Each level of consciousness and each body of spirit can appear only when all the previous ones are filled or evolved. Thus, the knowledge left by Nikolai Levashov allows us to obtain a complete picture of the structure of the human spirit and the connection between its material bodies and the levels of human consciousness.

# Chapter 7

# Knowledge from Ancient Egypt

The high priests of ancient Egypt knew the Laws of Spheres, and there is a lot of evidence for this. The myth of nine nested crystal spheres, which many people still hope to dig up in the sands of the Sahara, is not the only one to suggest that ancient people knew about the structure and laws of the space of spheres.

However, this knowledge was kept in the strictest secrecy for certain reasons, and it was possible to count all who possessed it on one's fingers. Humanity was given only a general idea of knowledge without revealing its true essence. It was "dressed" in an allegorical form, often personified as specific deities. In this chapter, you'll find several examples of it.

Having realized that the priests of ancient Egypt knew about the Laws of Spheres and many other things unknown to us, I decided to study the history of Egypt in-depth on the basis of publicly available materials. While exploring the works of scientists studying Egypt, both professionals and enthusiasts, I paid attention to the fact that many people who had visited Egypt once returned there again and again. Usually, they could not explain what force had been attracting and making them stay in this country. I am sure that this power is what any human soul feels unmistakably, and what has been lacking in many countries for so long. It was those invisible rays of love and sacred knowledge which were ignited by humankind many millennia ago and were not fully extinguished even during the millennia that had passed since this great ancient civilization was destroyed by other hostile and greedy nations.

The greatness and flourishing of the early period of ancient Egypt were explained by the fact that Egyptians granted power only to those people who had passed and withstood special severe tests to prove their fortitude and high level of consciousness. This practice guaranteed power to only the best of the best, as it should be in any reasonable civilization. Another factor was the natural life-affirming worldview that brought people together. It was the cult of life embodied in worshipping Ra (The Sun) and the superior laws of nature. Due to these two factors, Egypt became a prosperous and powerful country that no one could conquer.

Egypt did not fall overnight. It was a long, hard, and tragic battle between good and evil, which can be superficially perceived as a battle between supporters of the Ra cult and adherents of other religions. It was a brutal struggle accompanied by conspiracies, murders of pharaohs and their families, demolition of temples, the destruction of images and material evidence, the rewriting of history, and violent substitution of religions. Eventually, lies, meanness, betrayal, and direct military intervention brought the worst of the worst to power in Egypt, putting an end to the prosperity of this country.

The enemies of Egypt prioritized the destruction of the solar cult as an ideology that brought people together and made them strong. For example, once, as a substitute for the cult of Ra, a religious cult of the cat goddess was established in Egypt. Cats were declared to be sacred animals, and anyone who harmed them faced severe punishment. This religious and psychological trick was soon used by Hyksos. When attacking Egypt, their warriors had shields painted with the image of Bastet, the Egyptian cat goddess. They also used living cats to their advantage. Not wanting to offend the image of their goddess and sacred animal, the Egyptians were reluctant to fight and thus were easily conquered. This example demonstrates the importance of having the right ideology to ensure the survival of your civilization. Thus, the fall of Egypt as a civilization began with the violent substitution of the solar cult of life with other cults.

Egypt was most weakened by the introduction of the cult of Osiris. Undoubtedly, Egypt's enemies played a leading role in this ideological substitution. The same thing happened in other places, including Kievan Rus, where copies of this cult were introduced and adjusted to the authentic experience. Thus, the cult of Dionysius and Christianity substituted the natural ideas of life, freedom, and equality with the idea of slavery and tolerance, having paralyzed human consciousness. Tolerance is a medical term meaning a state of complete lack of immunity. In this state, any infection or parasites that have penetrated the body eventually destroy it. This is what generally happened to the "organisms" of great civilizations. The main idea of the new religion was as follows: "You are a slave (of God); be patient with everything that happens to you, as it is the will of God. You will be rewarded in the afterlife." However, the creators of this religion have "accidentally" forgotten to say the most important thing. The thing is that our consciousness determines our being not only on the physical level but also on other levels. Having left the physical body, a person gets into a world that fully corresponds to one's consciousness. If a person has accepted the idea of slavery and lived in this world according to it, then is there any reason to expect everything will radically change in another world? Surprisingly, many people do not understand such a simple and obvious thing.

Modern religions (starting with the cult of Osiris) were developed in a "genetic engineering laboratory," transforming the natural consciousness of people into one of slaves. The proof is that the malicious "gene" of slavery and obedience idea was implanted into a standard list of conventional natural principles, which have existed in the worldview of our ancestors for hundreds of thousands of years because they are the basic principles of life of any healthy civilization. After this implantation, religion turned into a sort of genetically modified corn. It looks good, tastes edible, but if you eat it, you will quickly die of cancer. Only, in this case, we are talking about the cancer of the soul, which is much more dangerous because you may have another body, but not soul.

Moreover, the mutation of consciousness caused by such religions is also fixed in human genes because it has been passed from generation to generation, leading to civilizational decay.

The insidiousness of this destructive weapon is that its results are not instantaneous but become visible only as generations change. Therefore, you should not be deceived by the psychological tricks of the religious developers, especially since they follow very different principles and ideologies. They live not as taught by their own religions. The visible variety of forms of these "genetically modified" religions should not deceive you either because all of them are equally dangerous for your consciousness.

It is actually quite easy to understand the essence of any such religion. To do this, we first need to determine whether this religion is tolerant of the parasitic social system. If it is tolerant, then it is most likely to be the product of this system. It is necessary to check whether the "genes" of slavery, sufferance, and tolerance are embedded in the religion. They destroy the natural immunity of a social organism. You also need to check whether this religion inspires the idea of superiority over others, with justification of the violence against them. Everything immediately falls into place after such a simple analysis.

The original worldview system, which has existed from time immemorial, is based on the worship of Ra, which brings life and light to humanity and has always rejected parasitism and slavery. Therefore, the forces of evil strived to destroy this worldview in the first place by substituting it with different types of slave religions. This happened not only in Egypt but also in other parts of the world, including the Great Tartary.

The purpose of this book is not to describe the true history of Egypt. However, speaking of this civilization, it is necessary to remember that knowledge and religion in late Egypt were very different from those that were in the early period of ancient Egypt founded about 13,000 years ago.

Here, I'm just giving some examples of the actual history so that you can better understand the reasons Egypt's knowledge was lost. I also want to give you an idea of how much the actual history has been distorted, perverted, and defamed by official "science" to conceal information about the events of the past. You must also understand the critical importance of knowing actual events of the past to draw the right conclusions about events of the present and correctly predict the future. Due to the official history, our consciousness is now stuffed with false information designed to keep us in the dark (i.e., in ignorance). Many truthful books are now banned in Russia for the same reason. Books can be destroyed or forbidden to print and distribute in an attempt to bring the Dark Ages back, but no one has the right and power to prevent you from reading them. No one can stop you if you want to know the truth that will set you free. You have been granted this right to know the truth by your very nature. Use it for your evolution, and don't let anyone take it away from you.

When I started reading about Egypt and studying pictures and drawings, there were two things that astounded me the most. The first is how insignificant and superficial is human knowledge about Egypt—the amazing oasis of culture, education, and understanding, despite the fact it played a significant role in developing Western, Near Eastern, and Middle Eastern civilizations. The second is that (it has not been a secret for scientists for a long time) most of the plots, characters, and ideas available in the Greek religion, as well as in modern Christianity, were simply copied from the Egyptian cult of Osiris. Christianity was composed mainly of transformed images and representations borrowed from Jews and Greeks. However, its foundations originate from the late period of ancient Egypt. The cult of Dionysius, as well as Christianity, accurately mimics the cult of Osiris, with only minor modifications of the plots [5]. In the Egyptian cult of Osiris, you will find both the god who created light and fashioned the first man out of clay, and the fact that god first created water and then separated the firmament from it. You will also find a divine horseman there, piercing the snake with his spear as it crawled on the ground, as well as the concept of a single god and God's judgment, the crucifixion and resurrection on the third day, baptism with water, and even the idea of an immaculate conception and a divine mother with a child in her arms that humanity began to worship not only in the countries surrounding Egypt but even in India and, as some scientists believe based on actual fact, in ancient America.

The ancient Egyptian civilization was not what humanity thought about it using intellect. It was a highly developed civilization in every respect. Egyptian hieroglyphs are found all over the world. They are present even in Australia. Most recently, I have learned that at least three settlements of Egyptians have been found near the Grand Canyon in the United States, with their common hieroglyphics on the walls of temples, as well as numerous artifacts. The place in the Grand Canyon was very similar to the canyons of Egypt, located along the Nile. This place was immediately labeled top secret, and the entire surrounding area was bought up by the military.

As ordered by high-ranking government officials, most of the discovered artifacts were gathered in a container and sunk in the ocean so as not to change the official version of the discovery of America by Columbus. Some of the discovered ancient devices were copied to create new military technology. In particular, an energy field generation device that had been found there was used to create field protection for tanks, making them invulnerable from the sides. Once, together with Drunvalo, I was near one of the Egyptian settlements near Sedona, Arizona. It was impossible not to feel the powerful energy present in this area. However, the government is extremely afraid of the truth. Governments are always afraid of the truth. Therefore, it is impossible to reach these places: they are guarded by lasers, and even planes are forbidden to fly above them. If you try to get closer, then an invisible helicopter will appear above your head, distinguishable only by the noise of the propeller.

Therefore, official versions should not be believed, especially in the field of history. We should seek knowledge ourselves, relying only on irrefutable evidence instead of works of official historians, whose purpose is not to find the truth but to hide it.

The deeper you dive into the history of Egypt, the more harmony you find in the life of the Egyptians. No, physically it was not easier for them; physical life has always been difficult. Nevertheless, they were a united people. Ancient Egyptians loved their land very much. They really loved it and left it very unwillingly, doing so only when they had to defend themselves against external envious and greedy peoples or to trade. The rulers of Egypt tried to help people and be equally fair to everybody, regardless of rank and position. The land was also divided equally. Geometry was born in Egypt, as a science of measuring land and dividing it equally among people. Egyptians believed that every person born in this land had a legitimate right to an equal share of the land and its gifts. So even the poorest man could count on his piece of land and, if necessary, on a fair trial and protection. The peculiarities of attitude to life, possession of knowledge, and natural worldview based on the worship of Ra (The Sun), who was shining equally for everybody and giving life to everybody, were the basis that gradually turned Egypt into a powerful and independent state, with wealth so extraordinary, the world composed legends about it.

The latter circumstance had apparently played a fatal role in the history of this amazing land. In any case, it was the main reason for Egypt becoming gradually filled with immigrants from other places, mainly from Palestine and other neighboring countries. Later, when sailing was established, there were lots of immigrants from Europe, mainly Greeks. It must be said that the Greeks and Romans had an extraordinary respect for Egypt and sent their students to study science and crafts there. The most prominent Greek scientists visited Egypt to find and acquire knowledge. Many Greek gods sitting on Olympus are directly related to Egyptian ones and look just like them.

In human imagination, Egypt is often associated with stone pyramids, which secured a large number of preserved material evidence and documented facts. If you look closely at the murals inside pyramids and preserved temples, you'll find traces of deeper knowledge present in the drawings, plots, and proportions, which was the priests' knowledge. Fig. 32 shows a fragment of a mural with an abstract from the Book of the Dead, telling the story of the path a person's soul will take after it leaves the world.

Fig. 32. Human life after death. A person needs to work hard to reach the stairway to heaven.

This picture shows a person facing another world, with five (in some cases six) spheres behind their back. These are the levels of consciousness. However, this person lacks four (three) levels to get to the stairs of the seven steps. The person is invited to earn them through their work in honest service to the gods. Further stories show this process of selfless service in a form accessible to the common folk. This process culminates in an opportunity to be admitted to the seven-stage ladder leading to heaven. You can see that this ladder is both an opportunity and a barrier on the way up. Whoever successfully passes it will be rewarded for their efforts and will rise, and people who will not stand the test will go down again. Any ladder can lead both up and down.

Drawing on a cosmetic box (ca. 1480 BC), Fig. 33 depicts the same knowledge, but expressed in a slightly different form. The unfinished Flower of Life, corresponding to the consciousness of man, is framed by five circles, denoting five levels of consciousness. Four larger circles symbolize the four levels of consciousness that man needs to develop to reach the completed planetary level of consciousness.

Fig. 33. The drawing showing the unfinished Flower of Life, corresponding to the five levels of human consciousness, and the four additional levels of consciousness that man needs to develop to complete his planetary cycle of development of consciousness.

Ancient civilizations knew a lot about the structure of consciousness and developed methods of transitioning to spaces of higher dimensions. The main corridor (Grand Gallery) of the Great Pyramid leading to the pharaoh's room (as scientists called this room, although it had nothing to do with the pharaohs), as well as the corridor inside most of the other pyramids, goes up precisely at an angle of 26.254 degrees — the angle leading to the fourth dimension. Many stone panels with drawings and texts found in the pyramids' rooms have been made by craftsmen within exact proportions of square roots of one and five.

The pyramids of the first three dynasties of pharaohs were very different from other pyramids; they were built ladder-shaped. During the Third Dynasty, the very first (at least among the preserved ones) 60-meter tall stone pyramid of Pharaoh Djoser was built. It was not actually a pyramid but a staircase going up high and leading to heaven. It consisted of six steps connecting seven levels, which were the seven spheres of the world. This symbol kickstarted the construction of the later pyramids. Later, it was assumed that a soul was to pass these seven levels after a person died. This general concept originated from Egypt, though it is present in all mythologies around the world.

The first time I truly understood that Egyptian priests had great knowledge was when I looked at the stepped pyramid of Djoser, Fig. 34 — not when I read legends of them. Imhotep was a brilliant architect, priest, astronomer, and, at the same time, vizier of Djoser. The Greeks equated him to the god of medicine Asclepius and recognized him as the greatest and wisest man, a leader of all researchers.

Imhotep was the first and only person recognized by humanity as a god while still alive. It's hard to even list all the areas where the genius of this extraordinary man was manifested, as well as all his merits, especially since there is not much information about him — even though the entire world once knew him. No one even knows where his body was buried or whether he died at all. If he died, he was wise enough not to make his body a target for robbers and hunters for mummies or let someone turn it into a piece of a museum exhibit, where he would be sullied by the fingers of passing tourists.

Fig. 34. The step pyramid of Djoser is a stairway leading to heaven.

In the movie *The Mummy*, Imhotep was portrayed as an evil magician, a monster who devoured people. (Hollywood has a long history of utilizing this type of imagery to conjure up an atmosphere of horror and suspense in the audience. It is a long-standing Biblical reference.) The reality could not be further from that idea. In reality, it was just the opposite. It's hard to find another person who has done so much for humankind. Imhotep was actually the founder of medicine. If humanity had not lost the knowledge of Egyptian hieroglyphics for such a long period of time, doctors would be giving an oath to Imhotep instead of the Hippocratic Oath. Greek scientists and doctors were visiting Egypt for centuries to learn about treatments, drug prescriptions, and descriptions of the properties of medical plants, which Imhotep left for humanity. During Imhotep's time, complex human body surgeries were performed, including operations on the skull. The honey bandage he invented on the basis of honey's properties can successfully replace modern antibiotics when laid on a wound.

However, this was only one of the areas of knowledge where Imhotep proved himself. He became known to the world also as a brilliant architect who started constructing stone buildings for the first time in human history (before that, humanity had used bricks for building). What he had invented in architecture was later used in other countries and is still used by architects today. The ladder-shaped pyramid he built in Sakkara was the first known stone structure. The very fact of how this pyramid was built speaks a lot. Before reaching the form we see now, it went through several stages roughly corresponding to the growth stages of the Tree of Life. At first, it was a very ancient stone mastaba, a large stone-lined parallelepiped with a square base. It was apparently left over from more ancient civilizations. It was extended by two smaller steps on two opposite sides, then another one was added on one side. The entire structure was eventually covered by a four-stage pyramid, with its steps connecting five levels. At the end, the final six-stage pyramid was built over this four-stage pyramid.

Take a look at Fig. 35. This pattern of blue faience tile is located over one of the passages between the rooms of the southern mastaba inside the pyramid of Djoser. Its proportions are also subject to the exact proportions of square roots. However, that's not all the designer of this amazing structure wanted to leave us.

Those who have eyes will see the Nine Crystal Spheres in this picture. There are nine levels of consciousness, five of which a man "holds in his hands" as the handle of the djed symbol this drawing is composed of. He has yet to develop another four. Humanity has lost its understanding of the djed symbol, which is often depicted in conjunction with the ankh, a symbol of eternal life, such as in a drawing of the Ani papyrus, Fig. 36 (left). Given that the circle has always symbolized God, and the ankh is a symbol of eternal life, you can now read the message that the author of this drawing wanted to convey to us. Sometimes the djed is depicted with the Flower of Life blossoming on its upper level, Fig. 36 (right). By all patterns, djed means human consciousness.

Fig. 35. The nine nested crystal spheres for which humanity has been looking for so long.

Fig. 36. The djed and ankh are often portrayed together (left). The Flower of Life sprouts from the upper level of the djed symbol (right).

The most important knowledge is concealed in the proportions of the steps of this pyramid. You may find a lot of secondary information about the pyramid — for example, how many million tons of stone was spent on construction, what kind of stone was used, what was the total length of underground corridors, and so on. However, I couldn't find the main one — heights of the steps. All I could find is their approximate size.

One book states that the steps were made with a gradually decreasing height: from 38 feet (the lower step) to 29 feet (the upper step). Another book demonstrates that the height of the steps was reduced from 22 qubits to 17 qubits, with each subsequent step smaller than the previous by one qubit (a qubit is a measure of length adopted in Egypt at that time, which equals 0.524 meters or 1.71818 feet).

None of these measurements can be considered accurate, so I'll give the steps' height values as they were originally in this pyramid. If the first step is 22 qubits exactly, these are the exact height values of the next five steps in qubits: 20.925, 19.993, 19.176, 18.452, and 17.804.

If the actual proportions are different from these, it is only because the pyramid surface has suffered damage over time. Some researchers have suggested that the pyramid should have had a seventh stage. Because this final stage has not survived, it is difficult to say what it looked like. Its exact height is 17.220 qubits. No one can explain why the height of the pyramid steps decreases bottom-up, and researchers can only make various assumptions. Thus, one of the scientists writes in his book that the steps were made in the supposed proportions of planetary spheres. In another book, the author writes that, when the architect built the first step, he probably became scared to have chosen too big an angle of the slope, so the pyramid could not withstand its own weight. Then he started decreasing the subsequent step heights to change this angle.

Reading various authors, I had the feeling that some of them, deep down, consider people who lived in ancient Egypt to be more primitive or stupid than we are. It is not true. They weren't smarter, but they were much wiser, for wisdom means not falling into the devil's service while serving science.

Those who possessed the knowledge took this knowledge with them without fully revealing it to humanity. They did it for our benefit, though we were separated from them by millennia. A modern scientist might find it difficult to understand such behavior. However, this test was not for those weak in spirit. They knew well that the knowledge of turning a desert into a flourishing oasis could quickly be used to turn an oasis into a lifeless desert if revealed to humanity. Knowing so much about Egypt, how could you assume that the high priest of Heliopolis, the founder of medicine, the leader of all astronomers, the vizier of the pharaoh, a man who was considered a wizard and god all over the world, who was worshipped as a god even by the first Christians, could be "frightened" by a simple error in his calculations? How would you explain why this "error" was repeated during the construction of other pyramids?

History has concealed the end of this story from us. Humanity believed that Imhotep's body was buried inside the same pyramid along with Djoser's body or beside it. However, no one has been able to find it. Faith in Imhotep and the sacredness of the ground at the foot of the pyramid was the reason why sufferers from various countries have been flocking to this place for many centuries, wanting and hoping to receive healing and help there. This sacred place was the last hope for many. People settled and lived right there, near the foot of the pyramid. Sometimes, the last thing they saw in this life was the ladder-shaped pyramid towering to the sky, the majestic ladder that took them to heaven forever.

Imhotep did not leave the Laws of Spheres for humanity. He knew dark times were coming. Still, he left his descendants with evidence that he possessed this knowledge, the evidence embodied in millions of tons of stone, a testimony that has survived through millennia. Only now can we appreciate the greatness of his act.

# Chapter 8

# The Tree of Life

What is the Tree of Life? Recently, this symbol has started frequently appearing in the literature. One day, I got intrigued by this symbol and started looking for an explanation for it. Having surfed through many books and internet pages, I did not find anything convincing. Nothing at all! It seemed that nobody knew what it was. If somebody does have the knowledge, they have kept it secret for some reason.

So let's do a little investigation together. Let's put aside dozens of new theories and assumptions about the Tree of Life. As you will see below, some of these theories have been simply invented by the intellect to hide the true knowledge of the Tree of Life (so that people would forget the way to immortality forever), and others were invented by the intellect simply because of its stupidity. Let's first see what the ancient sources say, as they are not so poisoned by intellect.

The Tree of Life symbol can be found in texts and drawings that are much older than Kabbalah and the Bible. As Drunvalo writes, the Tree of Life does not belong to the Egyptians either; it does not belong to any civilization or religion, or even a planet. Therefore, it's various images pop up everywhere on Earth. However, throughout history, these ancient images have gradually undergone changes and been interpreted differently by humanity at different times. In Genesis, the Tree of Life is described as a tree that bears fruit giving immortality to those who taste it. According to this story, after human beings tasted the fruit from the Tree of Knowledge, God banished them from the Garden of Eden. Since then, humanity has lost access to the Tree of Life and their immortality. That is, once we were immortal, and then we were turned mortal.

Sometimes, this tree is depicted with the serpent crawling on the ground around it, supposedly offering immortality (undoubtedly, those who interpret it in this way were convinced of this belief by the serpent itself), and with an eagle and a falcon sitting on its branches. In Egypt, such birds symbolized a spirit or soul capable of flying high into the sky.

In China, you can also find a similar tree, only with a dragon instead of a snake. I guess the Chinese like dragons more. Basically, it's the same, a reptile. In Egypt, it was a crocodile. The bird above was sometimes depicted next to the sun. Sometimes it held coins as a symbol of the true reward deserved by a soul. By the way, the Chinese Tree of Life is better defined: it bears a single fruit once every three thousand years. This fruit gives immortality to a single lucky person who finds it. This is not a comforting statistic for humanity, however.

In Arabic mythology, this beautiful, jeweled tree is located in paradise, next to the fountain of life. We can assume that, when reaching this tree, a person receives access to an energy source that makes one immortal.

In late Egyptian mythology, human beings originated from Isis and Osiris, both born from the Saosis acacia tree Egyptians considered to be the Tree of Life.

Aztecs also have the Tree of Life. It can be found on Russian icons that have undoubtedly borrowed it from more ancient sources. In general, the Tree of Life appears in epics, legends, and religions of various peoples around the world. One thing is certain: all these much similar interpretations and concepts have a single common source: the real knowledge of what the Tree of Life is.

Now let's abandon figurative and allegorical comparisons for a while and look at the geometric symbols usually used to depict this tree. I've picked these symbols from more or less credible sources. The most common symbols have come from sacred geometry, Fig. 37 (left), and Kabbalah, Fig. 37 (right).

They are similar, but there are some differences. You might wonder, so which of the two drawings is correct?

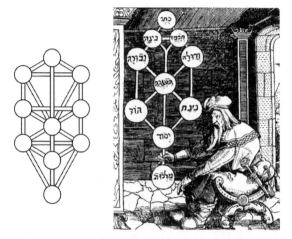

Fig. 37. Two drawings of the Tree of Life, which came to us from ancient times.

The answer is simple: both of them because they show us the same thing. To be mathematically accurate, the tree depicted in Fig. 37 (left) correctly reflects the geometric shapes we've already learned about. Identical distances between central spheres denoting layers are correctly observed in Fig. 37 (right). We won't be so scrupulous and picky. It's still hard to depict on paper what's happening in real multidimensionality. After all, the Tree of Life is simply a pattern of space deployment (including space of consciousness) in the sequence the spirit builds it. What we have drawn before is schematically depicted in these two figures. Here, the spheres are situated along the vertical central line; they separate spatial layers. The spheres to the right and left between each pair of central spheres indicate intermediate levels in respective layers.

If you look closely at these figures, you'll see the initial sphere the space of life begins with (because the mind is its direct attribute), a small tetrahedron, and an octahedron from

Fig. 4. You can also see that the number of intermediate levels grows with each new layer, similar to how tree branches grow. These numbers are accurately conveyed in the figure. This figure also shows two main shapes composing the space of life: a tetrahedron and an octahedron.

The Tree of Life is described in modern interpretations of the Kabbalah, mainly based upon drawings belonging to sacred geometry, Fig. 37 (left). Having scrolled through some of the books in Kabbalah, I noticed that the Tree of Life drawings had many arrows along the lines connecting the circles. Imagine how pleasantly surprised I was when, following their path, I found these arrows consistently connected all the spherical levels in the order they should have been connected!

Almost all books explain this path as something like the path of the divine energy descending into a human being. Unfortunately, that is the only resemblance. If you had only known that these fragments of secret knowledge were being piled up with all sorts of clever interpretations and conclusions belonging to a variety of authors! These Kabbalah interpreters have eventually confused and deceived themselves and others due to their implacable desire to appear wise.

If nature had been overthinking just like them when it was creating us, we would have probably been walking around curved, crooked, or humpbacked. We could have been walking upside down (I'm kidding). In its creation, everything is much simpler, more wonderful, and dizzyingly more beautiful. Its creation is perfect. If you practice sacred geometry, you will understand and sense it. Everything genius is really simple and perfect. It's just that sometimes it's not easy to comprehend the simplicity of superior harmony by such a "complex" intellect. Sure, I understand that it was not humans who did it and that all the time there has been someone behind them to whom it was...well, you know who I'm talking about. If you want, you can even trace the chronology of actions the devil has taken to destroy human knowledge of the Tree of Life, together with the Tree itself.

At first, the devil went down the most straightforward way, having turned into the serpent. He "repainted" all the circles in the Tree of Life from top to bottom and gave other names to each of them, choosing from a set of human qualities. He did it under a plausible pretext: ostensibly so that humanity would not forget about these qualities. He has just taken advantage of human trustfulness and naivety for his purposes. That was just the beginning of his far-reaching plans. Naturally, it didn't seem enough for him. Despite the fact that the majority of people would certainly believe this, there could be someone who would not trust blindly what one has been told and shown by others. This person could easily have spotted the errors and restored knowledge. Therefore, the serpent has started torturing the Tree of Life, taking advantage of human indifference and stupidity. For example, he put the fifth sphere between the four spheres of the upper layer, Fig. 38 (left), seemingly begging for symmetry on the vacant place, and called it "Knowledge." I do not know exactly what he meant by doing this. However, the author's hand is prominent here. Undoubtedly, this botanical experiment with the Tree of Life had been performed by the same one who persuaded humanity to deviate from the path of consciousness and go down the path of knowledge accumulation without understanding its nature. He has taken advantage of human recklessness and carelessness. With this name, the serpent exposed himself, but humanity didn't even notice it. Then he came up with another method. He planted the Tree of Life upside down, with its crown in the ground, Fig. 38 (right). However, the tree has still survived, despite everything.

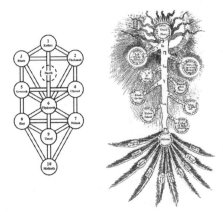

Fig. 38. Two examples of misconceptions about the Tree of Life expressed in drawings taken from different books.

Even after this seemingly irreversible distortion of nature in human representations, someone could have unraveled the true concept one day. Thus, the serpent couldn't sleep soundly.

That's why he proceeded even further. His brilliant project ended with implanting an error in the Tree of Life and multiplying it (that's what genetic "engineers" are doing now under his immediate guidance). Thus, the tree began resembling the serpent himself, Fig. 39 (left).

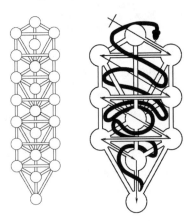

Fig. 39. Two other examples of misconceptions about the Tree of Life.

It was a real triumph. After such an invention, he could be completely sure that the tree had mutated beyond recognition, and its knowledge was lost forever. The serpent was delighted with himself. He triumphed and rejoiced. His ego and fantasy ran so wild that he could not resist the temptation to take a memorable picture. So he made his self-portrait among the Tree of Life branches with the cross, fallen in oblivion, in the background, Fig. 37 (right). He framed this portrait and hung it in the most prominent place of his office to mark another brilliant victory over humankind (one can imagine the fury of the serpent when he finds out I'm revealing the mysteries he's kept in the strictest secrecy for centuries).

Later on, the serpent had returned to this issue from time to time when he had nothing else to do, just to have fun and entertain himself with human stupidity. Thus, some book authors have developed "treatment" methods based upon Kabbalah, with his whispering in their ears. They have simply laid the drawing of the Tree of Life over a human body and begun assuring that the energy was penetrating a person exactly in the same ways that the spheres were connected. That is, the energy flow started from the head then reached the left and the right shoulders. From the right shoulder, it returned to the left one again, only slightly lower, and so on. They started treating people on the basis of this "mystical knowledge!" Some of these authors have degrees in medicine.

If other sections of the Kabbalah are interpreted in the same way (most likely they are), the best thing humanity can do in such a situation is to put aside tons of books dedicated to the "interpretation" of Kabbalah (so many living trees were ruined to destroy the Tree of Life!) and turn to the true knowledge that exists deep in the heart of each of us.

All this confusion about the Tree of Life has a simple reason. Humanity has completely forgotten the knowledge of the structure of consciousness and its laws. This is not surprising, considering that humanity has forgotten even what consciousness is.

The devil guided them to the path leading to death and oblivion. After this victory, he was sure that he'd put an end to the Tree of Life forever. However, he didn't take one crucial point into account. The roots of the Tree of Life go much deeper than the devil could penetrate. After all, he is secondary himself. What had a beginning has an end. The Tree of Life originates from the root source, from the point all human hearts contain. Now, humanity will have to remember everything. They have no choice because there is no other way to the future.

Considering and remembering this sad story, please understand that the knowledge we are talking about here is sacred. It is meant only for you, for those who have pure hearts. It's not meant for you to discuss with everyone and everywhere. It is for you to sense it within yourself the way you sense life itself. If you take this knowledge out, intellect will immediately take over. Then your knowledge will quickly become distorted. It will simply sink into speculation, as Kabbalah has been drowned in interpretations.

It looks like only Drunvalo has come close to understanding what the Tree of Life symbol means. Namely, he noticed that this drawing fit precisely into the Flower of Life and that the Tree of Life grew from the Seed of Life, which originated from the initial sphere the spirit used to start building space.

It's so easy to understand. You only need to adjust the order of lines connecting the spheres within the Tree of Life, Fig. 40 (left), and remember what is natural for the animate world. Like any living tree, the Tree of Life grows from a seed, which swells and turns into a sprout, which gradually grows into the Tree of Life.

Fig. 40 (left) shows two main shapes. There are only two of them, but they compose the entire growing space of life. This is the end of this tree. However, this is only the end of its "sprout" phase. The tree of spirit continues its growth. One day, there will come a time when a beautiful flower blossoms on the tree. All this tree requires is "suitable soil" and proper care, just like any other tree.

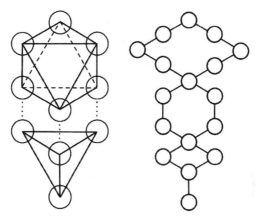

Fig. 40. Two interpretations of the growing Tree of Life.

Now, when you know how spaces work, we can further extend this tree, as shown in Fig. 40 (right). Here, the Tree of Life is presented not as a sequence (evolution) of the main shapes but as a map of layers and intermediate levels. Spheres that complete layers are placed vertically along the central line. As soon as the tree starts growing, it sprouts more and more new branches with each new layer, like any real tree. One day, it will bear fruit that will give immortality to those who eat it.

You see, as soon as you have shed some life-giving moisture of the consciousness on the soil that has been dried up by the dry winds of hot deserts of the intellect, the Tree of Life starts growing again. You have the same kind of tree; it's your own. Once you start looking after it carefully and lovingly, it will definitely show new shoots. It might give you fruit that will make you immortal in this life. It all depends on how you care for and nurture it with the living water of your consciousness. However, treating your Tree of Life with love and care is not enough; you still have to be extremely vigilant and attentive; otherwise, in the abnormal bustle around you, it can easily be trampled by others or...even by you. Don't forget that it's alive!

## Chapter 9

# Appearance of a Miraculous Icon

When the understanding of something very important came to me on my way to learning the truth, some significant confirmatory events took place in my life almost immediately afterward. Sometimes, it was a single event, and sometimes I experienced a series of them. However, all of them were quite extraordinary, unexpected, and unpredictable. The universe has countless, completely unpredictable ways of letting you know you're on the right path if you are open to understanding and noticing what is happening around you. I realized this a long time ago, so I've never even tried to expect or anticipate events. I have just watched them calmly and carefully while learning information and lessons for myself, being grateful for them. There were so many of these events that we could devote a second book to them. However, the description of this doesn't make much sense because each person has their own way, so I will only mention one of them.

After the Tree of Life revealed its secret to me, there was a significant event that I could not have imagined or dreamed up in my life. Angels (let's call them that) led me to a place where I ended up with a miraculous Russian icon in my hands. All this happened quite unusually, like in a dream, as if I was in two different states: at first, I was standing, and then I was standing with the icon in my hands. No one had given it to me; it appeared in my hands by itself.

You can imagine my surprise, given that all this was happening in the USA, in a place where, logically, it could not happen! I immediately saw that this was a real ancient Russian church icon. It had a large handmade trapezoidal hollow on a bog oak board on its back for mounting the icon on the church iconostasis. I was holding the icon in my hands while still unable to believe my eyes when suddenly a vision of a semi-dark room in an ancient Russian temple appeared in front of my eyes. It was illuminated only by the dim light of candles or icon lamps. I saw it all as if through the eyes of the icon! It was as if it had transferred this vision to me from its memory. In a single minute (though it seemed like eternity to me), a string of people passed in front of me. These people were suffering torments in their hearts (I felt them) and asking God for help and support in this extraordinarily cruel world. These people have probably been dead for a long time, but I saw them as if it was all happening in the present and as if they were coming up and asking me for help. My heart expanded with great compassion for all these people, for some of them could have been my ancestors. This feeling couldn't be conveyed by any words! I was standing there and looking at them. I didn't know how long it lasted because time ceased to exist for me at that moment. When I got back to the present, my eyes and cheeks were wet with tears.

I mentally thanked those who had presented this priceless gift to me and carefully brought the icon home. There, I took a closer look at it, and it fascinated me even more. I've seen numerous icons in my life, but I had never seen anything like this one. It did not have any symbols or scenes associated with the modern church or Biblical stories. The technique used by an unknown artist to paint it was different from all known to me. The edging ornament of the frame, the theme—everything was unusual. The icon, Fig. 41, depicted a "light" man in a long robe, standing quietly, with a pure blue river or stream smoothly flowing at his feet. I have used the word "light" rather than "holy" on purpose. In the Russian language, the word "holy" is simply a deliberately distorted equivalent of the word "light." It was done by churchmen who have always tried to lead people away from the light into the darkness of ignorance.

It was obvious that the person depicted on the icon was a man of light: a halo of light was depicted around his head. It was a glowing sphere. There was also an ancient temple behind him. In ancient times, before churches invaded the world, temples had been used to teach people knowledge. Surprisingly, there was a glowing sphere shining over the temple's dome, instead of a cross, which would be typical in modern Christian churches.

Fig. 41. An ancient Russian miraculous icon that suddenly landed in my hands.

The man's name was written in a rectangular hollow right over his head. Unfortunately, due to the icon's age and condition, the paint used for writing the letters had cracked and fallen away from the background, so that the name was impossible to read. Only the first capital, the old letter "P," could be read. It was weird that only his name had been erased. Everything else had been preserved relatively well. At the top, there was the Light Trinity, as if hovering in midair in the center of the icon. It was surrounded by another large transparent sphere placed exactly like the one on the famous icon of Andrey Rublev, but smaller in size. The icon was drawn with perfect technique. Despite the relatively small size, all the lines were flawless, and the face looked alive.

The most extraordinary thing about it was the sky stretching over the temple and the silhouette of a man. I've never seen anything like this before. It was not just golden. Before imposing a gilding layer, the artist had made a lot of voluminous hollows shaped like small circles and small radial lines on the ground, as if beams were emanating from them. So, when the sky shone with golden light, there were numerous little sparkling spheres. It looked like thousands of souls hovering in the sky, sparkling with golden halos. It was amazing!

Sculptured hollows were made almost everywhere on the ground. The artist covered them with paint, according to the theme and what they wanted to convey to people. The frame of the icon had a unique, slightly sculptured ornament.

Then, I noticed two deep, intersecting lines on the icon's surface. At first, I thought them to be scratches, but then I saw that the artist had made them on purpose. These two lines were like invisible threads connecting the heart and mind of a person with the Trinity, the superior spirits. It was as though the artist wanted to draw your attention to the state this person was in.

It was obvious that the artist had put their heart, mind, and soul into this icon. Since nothing happens by accident, the person's name on the icon might have been erased to demonstrate that it was not necessary to be excessively attached to names and personalities. Everyone can become this person of light, even you, if your desire is strong enough.

Following my revelation, I felt aftershocks of wonder for 3 days. For all that time, the angels were waiting patiently for my feelings to calm down. Then they told me calmly and briefly (they do everything quite calmly, very briefly, and comprehensibly) that the icon had arrived in my hands for a reason (I already understood that). They said the icon was stolen from a church in Central Russia many decades ago. Then it was bought by a man who moved overseas and lived in "the North." They didn't specify where exactly this North was: in Canada or in the north of the United States. They didn't specify how it ended up in the South, or how it came to me, but I didn't ask any of this. Now, I had a mission to return this icon to the people. They also said that I could keep it for a while or even forever, but that it would be better to return it to the people.

Surely, it would be better! I felt the need to return it to the nation and those people whose ancestors' faces had still been in front of my eyes. A new wave of feelings, awareness of my responsibility, and gratitude to the angels came over me for the following few days. How should I return it? I was in the United States, and I didn't know when I would visit Russia.

Sure, I could give it to, let's say, the Russian Orthodox Church in the USA, but deep inside, I didn't want to return it to any church. I wanted to return it to those Russian people it had been created for and then stolen from, along with their last hopes. I couldn't do anything about this desire. Mentally, I made various plans and pondered various options for how to implement this mission practically.

One must say that angels really have angelic patience. For 3 more days, they quietly watched my futile attempts to come up with a "plan to take the icon across the border," a "plan to give the icon to the Russian eparchy," and all sorts of other plans. Then they told me: "It will choose where it should be."

Angels always know how, with a single short phrase, to restore someone's sense of calm and equanimity. I calmed down immediately. I certainly had no doubt that the icon could do it! So, I just let things happen. I bought a closed frame with protective glass, put the icon inside, and hung it in the best place I could find in my house.

After that, I suddenly felt clearly that the angels had given me this icon not as a reward, and that it was not only the mission that mattered. A deep sense was concealed in this event and in this icon. It was a mystery for me to solve. I looked closely at the icon, trying to understand the knowledge concealed in it by the Superior Mind and revealed only to those who were able to embrace it.

This icon was a clue and a test at the same time. I felt that someone had been watching me invisibly, waiting for something. No doubt, this icon contained the key to understanding some very important, sacred truth. But what exactly was it? I carefully studied the details, trying to understand the relations between them, analyzing what I already knew. At one moment, it seemed to me that even the man on the icon smiled cunningly, waiting to see if I could solve the puzzle or not.

Three spheres of the Trinity were framed by one common larger sphere. It looked like a separated space of the Trinity. At the table where the Trinity sat, there should spatially have been a fourth sphere, which was not there for some reason. Who do you think it should belong to? To you. The place in the middle of the table with the offered delicacies is vacant and intended for you. You can come and join them. The Trinity has been waiting for you all this time and for so long: for centuries.

There were five steps to the temple, a large dome, two more small steps, and a small dome crowning it. Together, there were nine levels of the temple, standing on the ground and ending in the sky. Right over the small dome, there was no cross, which is usually depicted on modern icons, but another, fourth sphere. It was "floating" in midair, almost tangential to the three spheres of the Trinity. It was of a similar size, located on the same level but still outside their sphere…and a large green tree grew near the temple.

There was a pure river originating from a pure spring… pure spring…I felt it contained the key to understanding. In the Russian language, pure spring is also called "*kliutch*," which means the "key." You need to find a pure spring and "drink" only from it, to settle your life near it, excluding any impurity from what you "drink." It concerns informational impurity, impurity of thoughts and actions. After all, if you mix pure water with dirty water, then dirty water will not become pure, but there will be no pure, living water anymore. You need to get rid of all lies from your life—both big and small. It is not enough to get rid of them intellectually. You must get rid of them "organically," along with their source, which is poisoning pure, living waters.

Why was this knowledge revealed to me in relation to the Tree of Life? Probably because there was a large tree drawn on the icon near the temple by the river. While I was trying to understand the relationship between these two events, I had a vision of the Tree of Life, standing in front of the Living Water Fountain, as it had once been depicted in Arabic drawings.

Suddenly, the understanding dawned on me like a flash of lightning. I had revealed the main mystery of immortality!

You can become immortal only by evolving your consciousness (growing your Tree of Life), by acquiring knowledge and experience while drinking the water of information only from a completely pure spring that is not poisoned by lies!

You must drink from a pure spring and evolve your consciousness. That's the key to real immortality. The Tree of Life can grow when you are "drinking" information only from an absolutely pure spring that gives both life and immortality. Having realized and felt this knowledge and chosen to search for a pure spring, some people went to the desert, imprisoned themselves within the walls of monasteries and caves, or found solitude far away in the mountains. The Laws of Spheres say that the higher the place is, the higher and thinner its vibrations are. It is difficult to find and maintain spiritual purity while living among people, and therefore it is difficult to find immortality there. It is difficult to keep a spring pure in a place soaked in poisonous fumes of lies coming from people everywhere. However, people themselves are unlikely to understand this.

I am extremely grateful to the angels who revealed this true, priceless knowledge. After a while, another interesting thing happened for a slightly different reason. It turned out that soon afterward, I had to abandon my research due to numerous material problems. I had to switch to things that were purely "down-to-Earth" while solving the problem of my family's survival. This process took longer than I expected. It lasted for weeks and months. The angels waited patiently again, and then obviously decided to remind me of what was really important. One morning, when I was driving to work as usual in a slow and tedious stream of cars, a huge truck suddenly got up in front of me in my lane, blocking road visibility. Sure, there was nothing extraordinary in this fact itself. There were heavy trucks everywhere. For sure, you've seen a lot of them. However, I don't think there's anyone who's seen one like this: it had a green tree on it. A green tree was drawn on its huge backboard, with an eye-catching inscription made of large letters underneath it: "My goal is the Tree of Life." That was all! There were no other labels, no digits, no phone numbers—nothing that was usually present on trucks. It was so amazing that if I hadn't been holding the steering wheel of my car at the time, I would have thought it was just a dream.

This incident shook my consciousness and made me come out of the fog and stupor the human world had immersed me in. This incident finally assured me of what I had suspected for a long time while watching what was happening around me. Namely, I realized that events in our lives could be easily changed or shaped by someone who was invisibly watching us, and these events can take place in accordance with our condition and our goals.

You have no idea what power your angels have! The power capable of changing space, time, and events. It is a loving power that gives insight to those who really want to understand.

When I opened the icon's frame a few years later, I noticed a smoky silhouette of the depicted person and the temple imprinted on the glass. Churchmen usually present such facts as miracles. This icon stayed in my house for 5 years. Then it happened that it found a way to return back to its homeland, to Russia. It returned to Russia, and also directly into the hands of Nikolai Levashov.

If I describe here who Nicolai Levashov was and how he destroyed the devil's plans on this planet, you probably won't believe me—so unusual and extraordinary that person was. He was one of the most powerful hierarchs of the light forces. I do not know a single person on this planet with whom he could be compared. To truly understand who he was, you would have to spend some time and conduct your research. There is quite limited but enough information about him available in English. Evil people who don't want you to know the truth have many methods to deceive your intellect. They have used these methods for centuries. One of them is to defame or malign great people who reveal the truth. So, while doing your search, don't pay attention to poisoned sources of information, including Wikipedia. I would recommend you begin by visiting the *levashov.info* website and reading his biographical book *The Mirror of My Soul*. Then you will understand why the devil-made parasitic system hates him and why this icon landed in his hands.

# Chapter 10

# Space of the Golden Ratio

So far, we have looked at what sphere spaces are built around primary simplest shapes. Although the shapes above look simple, their real evolution was not as simple, nor was the evolution of various forms of consciousness. We have seen how one sphere turned into a tetrahedron and then transformed into an octahedron through informational complication and adding new dimensions.

Is that the limit? Let's try to find the answer to this question by looking at all these shapes at a slightly different angle. We can say that the point lies within the mathematical basis of one sphere. A tetrahedron is built based on an equilateral triangle and vertex points. An octahedron is built based on a regular quadrilateral (square) and includes equilateral triangles and vertex points. An evolution of objects is clearly discernible. Each subsequent, upgraded object (i.e., containing more dimensions) includes all previous ones. Understanding the evolution of Platonic solids is an important point, and many other relationships are difficult to understand without it.

We have already encountered the idea that a hexagon belongs to spaces based upon a tetrahedron and an octahedron. Therefore, following this logic, we still must consider only one mysterious shape—that is, a regular pentagon. It is "mysterious" due to its association with many secrets, legends, mystical teachings, and misunderstandings.

The regular pentagon itself is a truly mystical shape, at least because it is the shape of the golden ratio in its purest form. If you build a regular pentagon with sides equal to one, the distances from each of its vertices to the two opposite vertices will be exactly equal to the golden ratio $\Phi$ ($\Phi$ = 1.61803...). However, that's not all. If you study the pentagon deeper, you can see that everything touched by $\Phi$ (reads like "phi") acquires its "color, taste, and aroma" as if by magic. If you start connecting the vertices of the pentagon and measuring ratios of the segments inside it, you will get $\Phi$, or $\phi$ ($\phi = 1 / \Phi = 0.61803...$) everywhere. As you'll see later, this is true not only for the plane but also for three-dimensionality.

When I was exploring the golden ratio space, I always had the feeling that $\Phi$ was quietly laughing at us somewhere in its fourth dimension and wanting to play hide-and-seek. However, it is everywhere in this space, and each time it surprises you in a new way, manifesting itself in a completely unpredictable form and unexpected ratios. "Phi" is simply fascinating due to its magical properties. For example, $\Phi$ squared is equal to $\Phi$ plus one. $\Phi - 1 = \phi$, one minus $\phi$ squared equals to $\phi$, and $\Phi$ cubed divided into two is equal to $\Phi$ squared. Here's another interesting formula showing the very extraordinary properties of $\Phi$:

$$\Phi^x + \Phi^{x+1} = \Phi^{x+2}$$

These are not even all its phenomenal properties in nature, and we will explore these further in this chapter. However, if you start building space and layers around a regular pentagon under the same rules that we have applied so far, you will not obtain the desired result because this space is "magically" transformed by the $\Phi$ ratio. When I was researching various possible options for building "pentagonal" spaces, I always had a feeling that this was a kind of "parallel" space, where everything was controlled by $\Phi$ and which couldn't be accurately described either by whole numbers or by their square roots. It was penetrating our usual space and "soaking" it. As you will see below, the reality of the demonstrated "pentagon" can only be described by a mixture of all these values and their ratios.

If you start building the space of spheres based upon a pentagon, as we have done so far, you will not be able to keep all distances between the centers of spheres equal to one or to the square roots of the whole numbers. They must become equal to Φ or φ somewhere. Mathematically, though, there is a direct relationship between Φ and the sequence of square roots of whole numbers, Fig. 43. This figure, which can be extended indefinitely, mathematically, and visually shows "penetration" of the fourth dimension into our three-dimensionality.

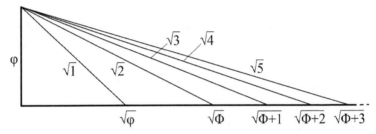

Fig. 42. Geometrical mapping of the relationship between Φ and φ and the sequence of square roots of whole numbers.

Sequences derived from the shapes where Φ is present are always formed by a mixture of square roots of whole numbers, Φ in different powers, and their ratios. For those who are interested in this topic, I have provided some more interesting facts in the "Additional information" section.

Well, now we know something that we can expect from Φ. Let's try to obtain a sequence belonging to "pentagonal" spaces together. Let's go from simple to complex, as usual.

Let's start with five adjoining spheres laid out in the form of a regular pentagon. To build a three-dimensional shape out of it, it seems necessary to put another sphere in its center (see Fig. 5) so that it touches the other five spheres. We'll obtain the so-called icosahedral cap. One sphere can also be added to the other side of the icosahedron. You will then obtain a shape consisting of two back-to-back icosahedral caps.

Let's see what sizes are concealed within this shape. Obviously, being the sphere diameter, there is a one (or root of one), as well as the value of Φ. We also need to find out what is the distance between the centers of the two spheres we have just added to the pentagon on both sides. Oddly enough, this distance is not equal to neither one, nor Φ, nor a square root of a whole number; it equals 1.05. This 5 percent deviation from one negates all attempts at building spaces using the known methods. No matter how hard we try to place spheres, there are "5 percent gaps" between them here and there. We can avoid these "gaps" only if we keep a distance between spheres equal to Φ.

This last circumstance is the reason for the laws the golden ratio space is based upon, which are quite different from the laws of previous spaces. Let's try to build space together by starting from the icosahedral cap.

Sure, it's possible to show geometrically how the gradual addition of spheres transforms the original shape, turning it into an icosahedron, Fig. 43, and then into a dodecahedron. As a result of this transformation, we can obtain only a wide sprawling shell formed by the spheres, which takes the correct symmetrical shape only at certain steps. We will not be able to obtain layers of the same thickness this way. Additionally, by adding a new sphere, we must shift the existing ones. It is also unclear what is inside and outside this shell. That's why there is not much sense in such constructions. This means that the space of a pentagon is built in a very different way. I was able to find only one way to build such a space, and this way is described below and depicted in Fig. 44.

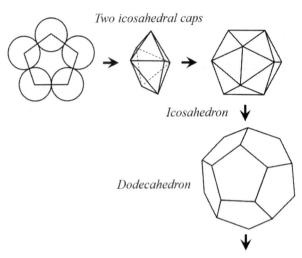

Fig. 43. The evolution of space based on the pentagon. Two icosahedral caps turn into an icosahedron and then into a dodecahedron.

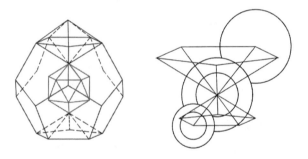

Fig. 44. Construction of space based on a pentagon: consistent alternation of icosahedrons and dodecahedrons. The spheres in each subsequent layer are Φ times larger than in the previous one.

To better understand this method, we must consider the "evolution" of the pentagon again. We can impose the same pentagon on the pentagon of spheres, having slightly turned it along the axis, so that its spheres lay exactly in the "gaps" between the spheres of the first pentagon. Then, one sphere can be added both to the top and bottom in the same way, Fig. 43. Two icosahedral caps are "dispersed" in this space, forming a shape called an icosahedron.

By measuring the distances between the centers of the spheres that form it, we can obtain a sequence consisting of five values:

$$1; \Phi; \Phi\sqrt{2}; \Phi^2; \Phi\sqrt{3}$$

or, which is the same,

$$\sqrt{1}; \sqrt{\Phi+1}; \sqrt{2(\Phi+1)}; \Phi+1; \sqrt{3(\Phi+1)})$$

Within this new sequence, two sequences are strangely intertwined and joined again — that is, a sequence of square roots of whole numbers and a sequence of the golden ratio. It is as if the two worlds, three-dimensional and four-dimensional (in their traditional sense), came together in our remarkable shape, bringing another dimension into our three-dimensionality. The important role of the icosahedron in this space should be noted. For example, from the ancient time this shape has long been associated with water.

Let me remind you that the sphere of the original diameter cannot be exactly embedded in the center of the icosahedron. It doesn't fit, even if we just put one pentagon on another without turning it around on its axis. It must become a little bigger in the diameter to fit in.

What's next? Then it turns out that you can build a second layer of spheres around the icosahedron and get the dodecahedron. However, this is only possible if the distances between the neighboring spheres of the next layer are equal not to one, but $\Phi$. In this case, the icosahedral cap of the next layer can be obtained by extending the sides of the previous layer cap with a simultaneous increase of the next layer spheres by $\Phi$

times, Fig. 44. That is, spheres grow in size with each new outer layer following the golden ratio.

That's not all. It also means that our upper sphere of a small icosahedral cap, Fig. 44 (right), is nested in the lower sphere of a large icosahedral cap (i.e., into a sphere of Φ times larger diameter!) Here, we learn about the concept of multilayered spheres for the first time. By continuing this construction up and down and taking into account the fact that we are dealing not with spheres, but with energies or levels of space, it turns out that each sphere consists of an infinite sequence of energy levels embedded in each other, and the entire space simply sparkles and shimmers with an endless variety of ordered multilayered "golden ratio" spheres. Each sphere is exactly tangential to the surfaces of the other spheres. What does it look like? It looks about the same as it is shown in the two-dimensional Fig. 45. In a voluminous form, the picture looks like a lot of spheres embedded in one another and next to one another.

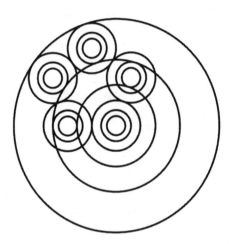

Fig. 45. This is how the space of the golden ratio looks in this section (the drawing shows only a small fragment).

The resulting dodecahedron adds a few more of its values to our range of distances between the centers of spheres, where various combinations of Φ and square roots of whole numbers are mixed.

Thus, the entire space is riddled with multilayered sliced spheres, with their diameters different by Φ times. Specific shapes in this space (icosahedrons and dodecahedrons) are manifested only if they are visually highlighted (or energetically activated). All this allows us to conclude that the space built based on a pentagon didn't "evolve" but always existed in this form.

From all this, we can conclude that there can indeed be an energy grid in the form of a dodecahedron above our planet, which Drunvalo has written about. It is usually called the "Christ Consciousness Grid." The grid, consisting of energy cells and connections between them, is an independent natural object related to the human consciousness—the evolution of our consciousness is impossible without it. However, in addition to this object, our planet also has its own consciousness with its structure defined by square roots of whole numbers.

Note that there is no both upper and lower limit in a sequence of these spheres. If we line up the golden ratio spheres so they are tangential to each other, as shown in Fig. 46 on the left, they will fit into the angle of 13.655...degrees.

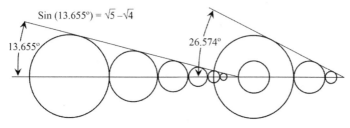

Fig. 46. Two specific angles of spheres of the golden ratio.

Don't ask why the angle is so strange. The answer is always the same: when we deal with the golden ratio space (i.e., we are immersed in the fourth dimension), we always deal with a square root of five in one form or another. Therefore, the square root of five or "Phi" derived from it are constantly manifested in this space in various unexpected ratios. That's why the sine of this angle is exactly equal to a square root of five minus the square root of four. Why is it the square root of four and not two? That's because we're looking at these spheres from the third dimension.

If you line up the spheres along the horizontal line not one by one, but every second sphere (see Fig. 46 on the right), this sequence of spheres will fit into the angle of the entrance to the fourth dimension, namely 26.565...degrees.

It is possible that separate specific levels of these multilayered spheres coincide with our square-root spheres manifested in the three-dimensionality. In other words, our manifested spheres of consciousness can be a special case of spheres of "divine" or "spatial" consciousness. This means that the square-root spheres inherent to us are created "in the image and likeness" of elusive "divine" spheres, which are also present in us invisibly. They help us connect with the entire world and all spaces. It can also be said that our consciousness is created "in the image and likeness of" and as a special case of the manifested "divine" consciousness.

## Chapter 11

# Squaring the Circle. True Value of Pi and the Great Pyramid

*Doubt is the first stage of knowledge.*
Michael Meyer, 1617

In this chapter we will discover the amazing knowledge concealed in squaring the circle and in the proportions of the Great Pyramid (the pyramid of Cheops).

One day Thoth asked Drunvalo to draw several concentric circles, with the gaps between them equal to the diameter of a small sphere. Then he told him to draw a square[1] around each of them and asked him to pay attention to the difference between the circumferences and the perimeters of squares that crossed them. This difference changed when moving from the center to the periphery, changing the sign, and increasingly approaching zero. Many people have heard and thus have an idea of the famous "squaring the circle" method, where the circumference length is equal to the perimeter of the square superimposed on it, Fig. 47.

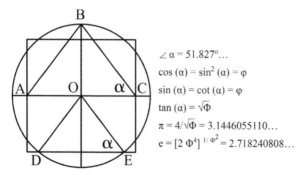

$\angle \alpha = 51.827° \ldots$

$\cos(\alpha) = \sin^2(\alpha) = \varphi$

$\sin(\alpha) = \cot(\alpha) = \varphi$

$\tan(\alpha) = \sqrt{\Phi}$

$\pi = 4/\sqrt{\Phi} = 3.1446055110\ldots$

$e = [2\,\Phi^4]^{1/\Phi^2} = 2.718240808\ldots$

Fig. 47. Squaring the circle reveals one of the mysteries of the Great Pyramid and also gives the true values of pi and e.

To not confuse it with another squaring, let's call it "squaring the circumference." There's another version of squaring of the circle, where the area of the circle is equal to the area of the square superimposed on it. Therefore, the term "squaring the circle" will be used in this latter case. Despite the uniqueness of these combinations of a circle (circumference) and a square, it isn't clear what's behind it and why all this is needed.

Drunvalo called the equality of the length of the circumference and the perimeter of the square the golden ratio. However, this is not entirely true because it is not the golden ratio, but simply a condition for equality of two lengths.

Because Drunvalo was doing all his constructions under Thoth's supervision, I realized that Thoth was telling Drunvalo something about the golden ratio in connection with its reflections on squaring the circle.

At one moment, I felt there had to be a direct and simple relationship between squaring the circle and the golden ratio. It's simple because everything in the spaces of spheres, including the golden ratio space of spheres, is wonderfully simple, interconnected, and perfect. I had already known that by that time, however. All I had to do was figure out the relationship. The simplest thing to begin with was to find a relationship between the circumference diameter and the square side where the known equality was performed. A circumference, as we know from school, is equal to the length of its diameter D, multiplied by pi (usually indicated by the Greek letter $\pi$) equal to 3.1415926...The perimeter of a square is equal to four lengths of its sides L—that is 4L. From the equality of these two values, we get our squaring the circle:

$$\frac{D}{L} = \frac{4}{\pi} = 1.2732395$$

When I first wrote down this ratio, the number on the right side seemed very familiar to me. I had surely come across it somewhere before in my calculations. I realized where exactly when I squared it. This number was almost equal to the square root of Φ. In the spaces of spheres, there is no such thing as "almost." Something doesn't match here, I thought. Something is wrong. Three out of four elements of this formula belonged to nature itself, and only one belonged to humans. There was no doubt what belonged to nature. Therefore, the only reason for this imperfection could be the value of pi wrongly chosen by humanity. So, if you write down our expression as:

$$\frac{4}{\pi} = \sqrt{\Phi}$$

Then you can derive the true value of $\pi$ from it:

$$\pi = \frac{4}{\sqrt{\Phi}} = 3.1446055110...$$

This number was about three thousandths bigger than the value of pi, as taught in school. This conclusion struck me so much that I decided to study the history of the origin of pi.

When I dug into this history more deeply, it turned out to be extremely similar to the history of adopted musical scales. It turned out that the value of pi was extremely different at different times for different nations. Its value varied from 3 to 3.5. Because no one could physically measure it directly, a variety of formulas were used to calculate it. For example, some people thought that $\pi$ was equal to the square root of ten. At present, two formulas for calculating $\pi$ are best known:

$$\frac{\pi}{4} = 1 - \frac{1}{3} + \frac{1}{5} + \frac{1}{7} + ...$$

$$\frac{\pi}{4} = arctan\left(\frac{1}{5}\right) - arctan\left(\frac{1}{239}\right)$$

Both formulas were obtained by mathematicians, but no one guaranteed that they were true, even though both of them resulted in the same number. For me, this circumstance was no less strange than for you. It was a serious step to claim that the pi value adopted a long time ago, rigidly encoded in the memory of millions of computers and calculators, inscribed in many computer softs and scientific books, was wrong. However, evidence was piling up that it was indeed wrong. On the basis of the formula I just calculated, it was impossible to say for sure that I had discovered the true value of $\pi$. However, it was the first "strike" against the dogma of the value of pi, priming it for collapse. I call it dogma because, as it turns out, it has no evidential backing. Nikolai Levashov once said that modern science is similar to religion because both are based on dogmas. At first, I didn't give this statement much importance; however, later, I have had many chances to become convinced of the genius of this comparison. The other thing that brings science and religion closer together is the most difficult and painful thing for both—that is, changing the adopted dogmas. For both of them, it is much easier to develop new dogmas (there have been more and more of them appearing over time) than change the old ones.

The second "strike" was the drawing of specific isosceles triangles inside the square and the circumference, as if they were connecting the circle and the square (see Fig. 47). When I drew both triangles, I suddenly had a vague feeling that this picture was familiar to me from long ago, as if it had come out of the depths of my memory. I felt that this drawing held the key to some important knowledge and contained some extraordinary harmony and interconnection. This feeling caused incomprehensible excitement. Everything after that happened almost by itself, without any effort on my part. I was just drawing and writing it down as if I'd known it for a long time.

When I calculated the inner angles of both triangles, it turned out that these triangles were almost similar. Again, I faced this "almost" which "did not count" in the geometry of the space of spheres. There was a deviation, though it was extremely small. The triangles were almost similar and were almost like a remarkable triangle from the space of spheres of the golden ratio, lying in the section of the Great Pyramid. This third "almost" was the last "strike" to the dogma of pi, resulting in this dogma finally collapsing, crumbling into small pieces.

I had put the legal, true value of pi on a pedestal. When I wiped off the thick layer of dust it had accumulated in the cellars of oblivion (by recalculating the parameters of triangles), then the new pi sparkled with dazzling golden brilliance. In this brilliance, it was surrounded by a halo of the radiance of golden spheres, and the square root of five, like the artwork of a brilliant artist, mesmerizing with perfect beauty, where it is impossible to find the slightest flaw. With the new value of pi, both triangles became absolutely similar to each other and like magic turned into the exact triangle of the golden ratio space, which was laid in the section of the Great Pyramid by the ancient Egyptians due to its extraordinary properties!

This triangle is really extraordinary. In it, the tilt angle $\angle\alpha = 51.827\ldots$ degrees.

The tangent and sine of this angle are equal to the following:

$$tan\alpha = \sqrt{\Phi}\,; sin\alpha = \frac{1}{\sqrt{\Phi}}$$

This is surely not all the extraordinary facts that can be found inside it. This absolute and perfect harmony is possible only with one single value of $\pi = 4/\sqrt{\Phi}$. It is simply amazing that the ancient Egyptians knew the true value of pi, having no computers or calculators. Modern scientists have spent months calculating the billions of digits after the decimal point while using incorrect formulas! This is not the last thing the Egyptians knew better than modern scientists. You certainly know a simple reason for this.

In the "Additional Information" section, you may find more information showing that the basic ratios between circles, circumferences, and squares, as well as between spheres and cubes, are determined by the golden ratio. That is, the geometric ratios that we have in our dimension came to us from a higher dimension, defined by the square root of five! This fourth dimension in our three-dimensional space is represented by geometric shapes of a circumference and a sphere, whereas our three-dimensionality itself can be represented by the shapes of a square and a cube.

The fact that squaring the circumference and squaring the circle are directly related to Φ, as well as to the inscribed and circumscribed circles and squares, confirms another interesting fact. Let's try squaring the circumference, circumscribing a new square around the circumference, and then imposing a circumference on this square. Together, they form the same squaring. If we repeat the same operation, Fig.48 (left), it will turn out that the diameters of circumferences (as well as sides of the squares) will be correlated with each other according to the golden ratio.

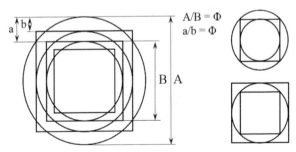

Fig. 48. "Magical" transformation of circumferences and squares in squaring the circumference and two specific shapes from a sequence of circumferences and squares.

The laws manifested in inscribed and circumscribed circumferences (spheres) and squares (cubes) suggest philosophical reflections on the ways and relations within the evolution of consciousness. After all, in the sequence of circumferences and squares, you can distinguish two specific shapes. You can distinguish a circumference between two squares and a square between two circumferences, Fig. 48 (right). You can see that the transition from a smaller square to a larger one is only possible through a circumference. Assuming that the square is connected with the specific consciousness of a living being manifested in the material world, this figure can be interpreted so that the evolution of a particular manifested consciousness of a living being (including humans) is carried out due to the influence of its "divine" aspect (the Higher Self or spirit), symbolically depicted in the form of a circle. Similarly, switching between levels of consciousness is possible only through the acquisition of experience in the manifested material world. That is, our consciousness and our spirit seem to go through evolution simultaneously and are inseparable from each other.

When I learned the truth about pi, I unwittingly asked myself a question. After all, it was so simple and perfect; why hasn't anyone figured this out for thousands of years?

I started surfing on the internet again and spent many hours trying to find any traces or hints that the truth had been revealed by someone. My search was successful. However, I found only a single person (after Plato) who had solved the problem of squaring the circumference and calculating the true value of pi. It was the Greek engineer Panagiotis Stefanides (*www.stefanides. gr*). While studying the works of Plato (and note Plato was known to have had obtained knowledge from Egyptians), Stefanides found that the translator had made a mistake while translating these works. As a result, "the most beautiful triangle," lying in the section of the Great Pyramid, was simply unnoticed by humanity! If the translator had not made the mistake, we would have another value of pi in school textbooks now. Who knows how many new ratios humankind has missed these many past centuries because of this simple translation mistake by someone who apparently knew little about mathematics or might have consciously concealed such important knowledge. It turned out that Plato, speaking of the "most beautiful triangle," wrote that this rectangular triangle had a hypotenuse equal to a cube and a large cathetus equal to a square of a small cathetus.

This is another wonderful property of the "most beautiful triangle!" The Pythagorean theorem can also be applied in this case. By mathematically exploring this triangle and applying the Euclid theorem, as well as using another method (i.e., solving the equation for the logarithmic spiral in the case when it becomes a circumference), Stefanides obtained a biquadratic equation of $\pi^4 + 4^2\pi^2 - 4^4 = 0$, with a solution of

$$\pi = \frac{4}{\sqrt{\Phi}} = 3.1446055110\ldots$$

The difference from traditional $\pi$ is about three thousandths, but this small difference allows you to reset all of the geometry, head to feet.

Then Stefanides found the true value of e (the basis of natural logarithm) because it was related to $\pi$. Look at the "ugly" formula e is expressed with in modern science because of the incorrect value of $\pi$:

$$e = \left[1.000924472\ldots\right]\left[2\left(\frac{4}{\pi}\right)^8\right]$$

By setting up the correct value of pi, Stefanides got the true value of e:

$$e = \left[ 2\Phi^4 \right]^{1/\Phi^2} = 2.718240808\ldots$$

Doesn't the latter formula look much more elegant? I exchanged several letters with the Greek engineer, so I have learned a little bit more about him. Stefanides discovered the true values of pi and e back in 1989. He immediately told the scientific world about his important finding, which resulted in a specific event in the scientific world. This event was...that absolutely nothing happened. As you might assume (otherwise I wouldn't have had to spend many hours searching for Stefanides' personal website in 2007), the scientific world completely ignored his reports and met them with cold silence and blatant sabotage. This is a typical experience of dealing with the scientific world, and I believe you have an idea of the true reasons for this silence: the reluctance to accept something new that goes radically beyond traditional scientific thinking and dogmas, the rejection by scientists of an engineer who started remaking science, the fear of losing their significance, ordinary human envy, or all taken together? It is difficult to say, but it showed once again that one should not expect recognition from scientists.

A confirmation of the above was the fact that, separately, scientists fully recognized the correctness of his formulas and calculations when meeting the Greek engineer: they simply had no arguments to refute the obvious truths and correct formulas. However, when it came to public recognition, they seemed to immediately forget about the existence of Stefanides and his work. Everything proves that modern scientists, at least most of them, do not need truth at all, and science has turned into a religion based upon dogmas. Our finding has just easily broken two of them (i.e., dogmas) about pi and e. The situation isn't better, with other dogmas lying in the foundation of modern science as well.

In the scientific world, as well as in many other areas, something new is usually recognized only under the influence of "magical" influencers or mass advertising, even if this new thing is complete stupidity, delusion, or lies. However, both "influencers" and advertising are in no hurry to surround new talented people with attention. Actually, it is just the opposite. An artist, Henry Fuller, has unusually accurately depicted what was happening in science on his canvas, Fig. 49. This painting will help you cast off one of the strongest and most enduring illusions—namely, of trust in modern scientists and science. Humankind is shown on this canvas in the form of a naive child reaching for science to gain the knowledge it needs for development. But it receives only bubbles of illusions instead of knowledge. Science is depicted as a woman living in a luxurious building and dressed in expensive clothes, which, however, do not hide her nakedness. As a result, both humankind and science remain in complete ignorance of nature and its laws, which is symbolized by nature in the painting background, immersed in the darkness of night.

Fig. 49. Henry Fuller's painting "Illusions" can be considered as a symbol of modern science, as well as religion, education, media, medicine, and other products of the parasitic social system, created to keep people ignorant.

As if confirming this fact, Russian scientist Yuri Rybnikov showed that modern mathematics is a completely abstract science that has no connection to the real world and is based on false mathematical operations and concepts. In nature, there are only two operations: summing and subtraction. All other operations are based on these two. What did the scientists do? They replaced summing with the term "addition" (in the Russian language, this word is based on the root "lie") and the multiplication operation was called "exponentiation." In nature, when living cells breed, they multiply. Scientists, for some reason, have called this process division, having turned everything upside down. Even these simple examples show that false concepts have been introduced into the very foundations of science, confusing minds and leading humankind away from understanding the real values and laws of nature.

Rybnikov showed that modern chemistry is also based on a false table of elements and false ideas about the laws of interaction between them. The correct table, compiled by Mendeleev, based on ancient knowledge known to our ancestors, was "genetically modified" on purpose, so that humanity would never discover the real knowledge and method of obtaining free energy.

The mathematics of our ancestors has always been associated with real natural objects. For example, if you have three apples, you can't subtract more than three apples from them. If you subtract six from three, following modern mathematics, you will obtain minus three apples, which does not make any sense in nature. Negative numbers were introduced into mathematics only with the advent of usury. Even Wikipedia admits it. Moneylenders and bankers needed a "science" that justified taking away other people's property by borrowing money in growth and creating debts—ultimately justifying enslavement. These actions are completely alien to the laws of nature. According to the law of mutual assistance given to us by nature, when you voluntarily give something to a person worthy of such help, you cannot demand that they return it to you, let alone that they return more than you had given. In real life, people may not be able to return what you voluntarily gave them. If they have the ability, they will return your kindness. In nature, for example, when a wolf brings a piece of prey from the hunt to a sick member of the flock, it does not require its relative to repay the debt or return a larger piece of meat in the future. However, our intellect is now so clogged with false notions of the false relationships between people that it is already difficult for many of us to understand even such simple, natural things.

By using their own "mathematics," bankers have solved our matter in their own way. They came to you and said, "We're taking away your three apples, and you're going to owe us three more apples because they have a "minus" sign. If you don't have them, you'll work for us until you give them back." Of course, you don't have them! How can you have something that doesn't exist in nature!? The most interesting thing is that humanity believed the bankers and actually started working for them to give back what they should not give back at all! They should have answered the bankers:

"Working for you means going against the laws of nature— and therefore against ourselves. We're not going to do that. For your criminal thinking and actions, we'll send you to prison for rehabilitation, where you'll have enough time to reflect on the laws of nature." That's what Icelanders once did to their bankers. Why don't people from other countries do the same? Probably because they do not know the laws of nature well and do not deal with them every day, as do the inhabitants of that remote northern country.

A. P. Kiselev, in an arithmetic textbook used by several generations of people to study a hundred years ago and before, stated that, "from any number it is impossible to subtract a number that is greater than it. Therefore, the subtrahend cannot be greater than the minuend. No number can be subtracted from zero because all other numbers are greater than zero." Modern scientists have "genetically modified" mathematics (do you recognize the hand?) by introducing elements that do not belong to it at all. For example, the terms "zero" and "infinity" are not numbers at all. That's why we can't do any calculations with them. A zero is simply a concept that means an absence of any material object or any operations over them. Infinity means knowledge uncertainty on a scale with undefined limits. Therefore, it is impossible, for example, to sum something with zero or infinity or to multiply or divide by zero or infinity. If you have an apple and you multiply it by zero, then you will have the same apple. Mathematicians have confiscated this apple from you with the help of an operation by turning it to zero. However, subtracting larger numbers from smaller numbers (your debt to banks) and multiplying by zero (withdrawal of your savings) or by infinity (promising you something in the uncertain distant future with the absence of it in the present) are just perfect operations for bankers and those in power. Modern "science" has simply served bankers' needs for money by manipulating concepts and distorting the understanding of reality and its laws for humanity and presenting illusions and false dogmata in return. For the same reason, real scientists who discover knowledge that can set humankind free have been persecuted and physically exterminated for thousands of years, along with the knowledge they had discovered.

As you can see from the facts, modern science and religion are not only similar to each other, like two sisters, but both of them are the product of the same "parent" — that is, a parasitic social system. Both of them serve its interests and are therefore supported by it. Here we have finally come to realize that the genius artist Henry Fuller depicted not only the science but also the entire parasitic system as a whole on his canvas. Therefore, the scene of Henry Fuller's painting can be equally attributed to science, education, religion, politics, medicine, media, economics, and other areas of the system. From these examples, it becomes clear how important it is to understand the laws of nature on your own, without relying on any influencers and scientists.

If we had not dug deeper into the issue of the relationships between mathematics and nature, we would not have found the lie concealed in it, alongside the lie in the system that gave rise to mathematics and other sciences. We would have lived in the illusion for the rest of our lives, thinking that we owe something to bankers and that science itself would make human lives better. It won't. Only your consciousness evolution through learning the laws of nature—and true understanding of these laws—can make life better.

Considering these particular qualities of modern scientists, you can quite confidently use the new values of two important constants and new, absolutely accurate geometric formulas, without concern over the opinions of scientists. For your convenience, I have put these formulas into a table and placed them at the end of the book.

Even though the way Stefanides calculated the true value of pi was too difficult (like most modern scientific methods), and the origin and value of this constant remained unsolved by him, nevertheless his work was strong proof of what I have come to understand in very different ways.

As you can see, both pi and e take their origin from Phi, the golden ratio. At the heart of it is the square root of five, the value that is the unit of the fourth dimension, setting the ratios in our three-dimensionality. This is amazing!

Formulas are only an abstract representation of what exists in reality. In the equation

$$\Phi = \frac{\sqrt{5}+1}{2}$$

there is clearly some real value concealed. The fact is that, in the space of square roots, real spheres have actually a diameter that is twice larger than that of our abstract mathematical spheres (which have empty gaps between them). In nature, as we know, there is no emptiness.

Undoubtedly, Φ directly originates from the square root of five, which is a unit of the fourth dimension. If we depict the real sphere, Fig. 50, instead of an abstract one, with twice as large a diameter, and its center distanced from the center at a square root of five, and we measure the distance from the center of the cluster to its outer active shell (which determines the size of the entire body in the fourth dimension), then we will get a distance equal to $\sqrt{5} + 1$. This works only in cases where we look at the spheres from the very first (zero) dimension. In fact, we are in the third dimension, from where everything "seems" exactly half as big.

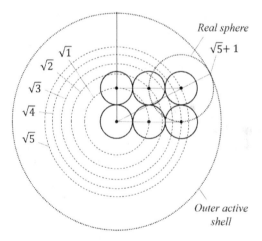

Fig. 50. A drawing that gives you a visual idea of the origin of the Φ value in the space of the square roots: the real spheres and the active outer shell.

Taking into account this correction, the outer boundaries of the shells of real clusters of spheres will be determined by a slightly different sequence:

$$\frac{\sqrt{1}+1}{2}; \frac{\sqrt{2}+1}{2}; \frac{\sqrt{3}+1}{2}; \frac{\sqrt{4}+1}{2}; \frac{\sqrt{5}+1}{2};...$$

Φ in its pure form stands in the fifth place of this sequence. That is, we can say that Φ determines the boundaries of a body (cluster of spheres) from the fourth dimension. This size is set not by Φ, but by the square root of five. It is from this root that many (if not all) constants have come to us, including pi and e, and this fact itself is just amazing. It is also striking that scientists still do not want to admit it. We can assume that if you were in the fourth dimension (I mean, if you were there as a whole, because a part of you is already there), then for you the value of Φ would be the value we perceive from our dimension as

$$\frac{\sqrt{6}+1}{2} = 1.72474487...$$

Because the mathematics of space built on the basis of a pentagon is completely different from the already familiar mathematics of the previous three spaces; it may be doubtful whether such a "pentagonal" space of spheres is capable of existing, and is it really present in nature? Plato and other researchers of the sacred geometry attached a special, priority importance to the figures we have just considered. Drunvalo wrote that, "In the Pythagorean school, if you even uttered the word 'dodecahedron' outside the school, they would kill you on the spot."

What many people have kept as a sanctity at different times is hardly based on nothing. According to Drunvalo, "The dodecahedron/icosahedral relationship is near the outer edge in your energy field and is the highest form of consciousness. On a microscopic level, the dodecahedron and the icosahedron are the relational parameters of DNA, the blueprint of all life."

Nevertheless, we cannot say now that we have completely revealed the mystery concealed in the pentagon. Maybe this is just because it's not the time for humanity to know it. Perhaps, we simply cannot comprehend it at our level of consciousness. Apparently, it hides a certain "divine" or "universal" value. It's possible that you will be able to discover some of these mysteries personally. Regardless, you have already learned quite a lot.

## Endnote

1. When I made a similar construction with the circumferences of the space of square roots of whole numbers, I obtained a phenomenally accurate matching of squares to circles. The square drawn around one circle appeared to be precisely inscribed in another one of a larger diameter. The areas of these squares were correlated as whole numbers 1, 2, 3 . . .

# Chapter 12

# The Main Mystery of the Great Pyramid

The more drawings I made, the more and more secrets of the "golden ratio" spheres were revealed to me. At times, it seemed to me that all the laws of the world were concealed among them. In the spheres, I have found not only the ankh but also the Great Pyramid, and the main known constants, and other symbols, and many, many amazing relations which are difficult to tell about in one book. I have also realized that many of the incomprehensible circles and drawings that other civilizations have left in the fields were nothing more than spatial maps and encodings. I have managed to decode some of them, but I want to let humanity learn it for themselves. If humankind had not voluntarily given all power to the crazy minds of bankers, politicians, and military officers, putting troops on alert and chasing flying saucers to obtain new technologies for military purposes and ultimately to further enslave humanity, if they had carefully studied the messages and evolved their minds, they would have learned to travel through dimensions and spaces long ago.

The Superior Mind sends these drawings so that we can evolve ourselves and learn to feel the love and magical harmony of the laws from which this world was created. We must not destroy this harmony. There are aliens hostile to humanity, such as the "greys," who kidnap people and animals to experiment on them, who are able to lie, feel envious, and fail to keep promises. These inner qualities bring them closer to politicians and make it possible for them to cooperate with governments. They are rather an exception to the general rule though.

You may be unable to understand all sorts of spatial drawings, but still build such a harmonious and happy life around you, that those civilizations, which are good at making and reading such drawings, will come to you to learn from your experience.

Drawings and flying saucers have not made anyone happy yet. After all, you were born on Earth; here there is everything you need for your life, and there is no need to fly to other planets seeking happiness. If you don't find it here, you won't find it there either.

Yet the golden ratio space contains such perfect mathematics that I can't help but share some of my research results with you, so that you can enjoy the perfect harmony of its laws too.

If you symbolically arrange the spheres of the golden ratio sequence as they are located in spatial layers in a figure, it will turn into a rather curious one. In Fig. 51, these circles are highlighted in bold line. The transition from sphere to sphere occurs as if through a sequence of eights "penetrating" each other in $\Phi$ proportion, coming from some starting point (in the center), but going from quantitative infinity to also infinity. You can see that with the diameter of Sphere 1 in Fig. 51, conventionally equal to one, the convergence point of a sequence of spheres built in space will be remote from the center of Sphere 1 to a distance equal to $\Phi^2$. Everything is surprisingly interconnected within this sequence.

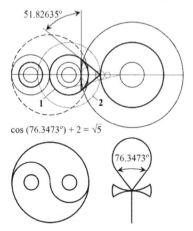

Fig. 51. Spheres of the golden ratio space shaping, the triangle of the Great Pyramid, and the symbols of ankh and yin-yang.

In this figure, you can see the symbol of the ankh and even the yin-yang symbol. Curiously, the circle diameters in both symbols obtained in this way encode the angle of transition to the fourth dimension, namely the angle of 26.565…degrees.

In his "Emerald Tablets," Thoth said that he had built the Great Pyramid using the proportions of the Earth and the Moon. In Fig. 52, you can see three pairs of circles of the golden ratio sequence (one pair is shown on the left and the other one is on the right), and the diameter's ratio exactly equals the ratio of the diameters of the Earth and the Moon. This illustrates that the moon is an artificial object created to harmonize space around the Earth. Actually, there were three moons in the past; two of them were destroyed a long time ago. The second one was destroyed about 13,000 years ago. Some of its pieces impacted the Earth, resulting in what is commonly known as the Great Flood. Unfortunately, the space wars are not a fantasy but reality.

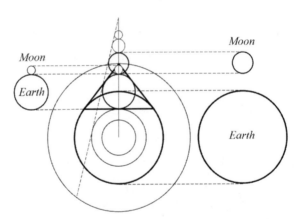

Fig. 52. Proportions of the Earth and the Moon concealed in the Great Pyramid.

Perhaps the most interesting point in our construction is the definition of a visual angle at which point each sphere of the golden ratio space is visible from the surface of the next sphere of a larger diameter. Imagine that we are on the surface of some sphere, looking at the sphere of the previous level; for example, we are looking at Sphere 1 from the surface of Sphere 2. This sphere is visible at an angle of 76.347 degrees. That is, if we took two tangents from any point of Sphere 2 to Sphere 1, the angle between these tangents will be 76.347 degrees. Now let's connect the tangency points using a straight line segment. In the resulting isosceles triangle (it is highlighted on the figure in bold line), the tilt angles of the sides are equal to 56.826 degrees. Thus, the remarkable triangle of the Great Pyramid appears in front of us again! It is as if it has reappeared to invite us to follow it into the Great Pyramid itself and reveal the secrets concealed in the depths of millennia. Well, let's finally replace our magical triangle with an image of the Great Pyramid section and see where the most important points of the real pyramid will be situated in the space of the golden ratio spheres, Fig. 53.

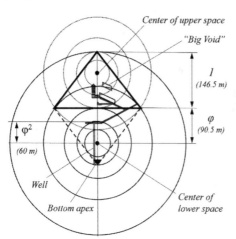

Fig. 53. The Great Pyramid binds together several levels of the space of spheres of the golden ratio, whose center is located at a depth of about 60 meters from the level of the subterranean chamber. A mysterious backfilled well goes down to the lower vertex of the pyramid.

It is clearly visible that the pyramid seems to connect several golden ratio spheres—that is, it is like a resonator or a crystal portal connecting several dimensions in the golden ratio sphere space.

To better understand what this figure says, you need to remember some information about the Great Pyramid. Both main corridors of the pyramid, the one rising up from the main entrance and the other going down, go exactly at the angle of transition to the fourth dimension.

Despite the prohibition, one visitor went up to the apex of the pyramid and wrote that, as soon as he reached the top, he felt a powerful flow of energy causing him to collapse. He was forced to immediately go a few steps down.

Drunvalo wrote that, when he was descending into the bottom corridor, which is closed to tourists now, at some point he clearly felt a leaping drop in the vibrational frequency of the space—an entire octave. This happened twice as he was walking down the corridor. The spots where it happened were marked with two red squares on the walls. The second place was at the entrance to the subterranean chamber. In the subterranean chamber of the pyramid, there was a backfilled well made for an unknown purpose, leading to an unknown place. Above the pharaoh's room, there was a resonator—that is, a set of polished stone mirrors located atop of each other, complete with two massive reflecting stone slabs at the top. Thoth also told Drunvalo that the pyramid was built with exactly the same pyramid underneath (i.e., a pyramid placed over a pyramid).

All this information is perfectly consistent with the map of the golden ratio space we have obtained, which is linked by the Great Pyramid. You can see that one of the spheres crosses the subterranean chamber at its beginning, passes along its ceiling, and then crosses the chamber level again at the end of the dug-out solid tunnel. This is the very spot

Drunvalo and his companions initiated as they were guided by Thoth. At this place, Drunvalo claimed to be exposed to dark light energy leading to the Halls of Amenti, located in the bowels of the Earth. When you are at this place, your thoughts may become reality. Some people disappeared in this place, and they were then found in a sarcophagus in the pharaoh's room. That's why the sarcophagus was moved from its original location. The other sphere is precisely tangential to the base of the pyramid. This pyramid seems to lean on the next sphere. This sphere also passes precisely through the top of the stone resonator, well-described by Drunvalo as an acoustic generator. According to Drunvalo, the resonator is located above the pharaoh's room, where one experiences the white light energy. The apex of the pyramid falls precisely to the surface of the largest sphere.

We can say even more about the Great Pyramid. Thoth told Drunvalo that he had started building a pyramid from its vertex. Without seeing Fig. 53, this phrase seems to make no sense. However, he actually started building the pyramid from its vertex — that is, the vertex of the bottom pyramid. It is most likely that Thoth put an energy crystal in there. Therefore, the mysterious well in the subterranean chamber of the pyramid goes to its lower vertex. It is also possible that something very important is placed at a depth of just under 60 meters. The space of the golden ratio spheres begins at this spot, with its levels connected by the Great Pyramid. There is also another space of the golden ratio spheres that begins at a similar point of the upper pyramid (these spheres give us another set of key points). According to the space structure, there may be another corridor above the main corridor (Grand Gallery) in the place connecting the first and the second level of the upper space of spheres. That corridor is shown in Fig. 53. Recently, scientists using cosmic-ray muon radiography discovered an emptiness they called the "Big Void," which is located exactly at that place.

Most likely, an energy field or harmonizing space crystals are placed at the centers of both spaces of spheres. There also might be an entrance to another space or to the Halls of Amenti. However, this mystery will only be revealed to humanity in the future.

# Epilogue

In this book, you have been exposed to knowledge that was concealed in sacred geometry, and the key to its understanding was lost by humanity thousands of years ago.

Unfortunately, people with unevolved or distorted consciousness, who have seized power on this planet, do everything to slow down the evolution of human consciousness and create obstacles to the natural laws of evolution. The rulers want to keep humanity in darkness and ignorance to gain control over human consciousness, which are necessary conditions for maintaining their power. The sculpture, whose fragments are shown in Fig. 54 (similar sculptures can be found in various countries around the world), can be called a symbol of parasitic power. All that the parasitic social system has to offer humankind is the suppression of the natural process of human consciousness evolution and primitive slavery. It's incapable of anything more, and this sculpture clearly demonstrates that.

Fig. 54. The main condition of the existence of a parasitic system is control over human consciousness. It is reflected in this fragment of the sculpture of a lion having put its paw over the sphere of human consciousness.

The methods used for this are well known. Everything is put into play, from the media (which have actually been turned into a tool of lies and misinformation), distorted morality, ideologies separating humankind, black magic, psy-weapons, murdering the best of the best, and to the genocide of entire nations. Of course, everything has been done to establish and maintain the power of the worst of the worst. Because human consciousness is directly connected to the consciousness of the planet, these people also do everything to bring down the consciousness of our planet to a lower level as well.

Take a look at the photograph in Fig. 55, and you will see how these people are trying to burn the levels of the planet's consciousness with the help of nuclear explosions, which are causing irreparable harm to all living beings, including themselves. They are so blind in their thirst for power that they do not even pay attention to it. In the photo, you can see how a nuclear explosion discloses the boundaries of the levels of the planet's consciousness, which are subject to a sequence of square roots. Fortunately, the laws of evolution are unstoppable, and the parasitic system of human relations will come to an end, sooner or later. Many wonderful people often sacrifice their lives, as they do everything, to bring this time closer and create the conditions for the normal evolution of consciousness of humankind living on our planet. There is only one sure-fire way to resist all the artificially created factors that destroy consciousness—that is, knowing the truth and evolving your own consciousness.

In the book *Serpent of Light*, Drunvalo described the process where the light beam output traveled on the surface of the Earth from India, where it had stayed for several thousand years, to northern Chile. He also spoke about the process of transitioning the Earth to the fourth dimension. Many people misinterpreted this information simply because they lacked information to understand this process. Some of them even decided that all people must die in the third dimension to "resurrect" in the fourth one during such a transition. However, we are talking only about transitioning the planet's consciousness to a new level, as you have probably already guessed, which will undoubtedly have a positive impact on the evolution of the consciousness of all people living on it.

Fig. 55. Nuclear explosions in sacred places of the planet are the government's secret attempts to reduce the consciousness of the planet and humanity to a lower level.

Thoth once told Drunvalo that the level of our planet's consciousness corresponds to the tilt angle of its axis. During the planetary catastrophe and the offset of the Earth's rotation axis, which occurred about 13,000 years ago as a result of the war with Atlantis and the destruction of the second moon, this level was pushed down. However, now the Earth's Kundalini (as Drunvalo called it) has changed its location on the planet's surface again. Let's see if this event will in any way affect the planet's consciousness.

To answer this question, let's look at the latitude of northern Chile. This latitude is exactly equal to the 26.565-degree angle, which is exactly the angle of entry to the fourth dimension. It can be assumed that the place where the Earth's Kundalini was in India before corresponded to the latitude of 30 degrees. That is, the Earth already moved to the fourth dimension at the moment when the Kundalini beam had completed its journey from India to northern Chile. So now it's just humanity's turn.

From this book, you have learned some of the important laws that were born together with the world. These laws can be applied to practical use, just as they have been used by ancient civilizations to encourage human consciousness to move to new, higher levels.

One possible application of these laws is the construction of a new structure type, which can be called the Temple of Divine Spheres. This new temple will not have any religion associated with it. It will simply embody all the knowledge you have learned, as well as things you don't know yet. In this building, everything will be made based on the laws of life and the mind: from general proportions to internal perspective, sound, color, and light solutions. By immersing oneself in natural proportions of sound and light in this temple, the soul of each person will be able to feel a divine experience and rise to new, divine heights. Isn't that the experience we live for? Whether such a structure will be built depends on the willingness of humanity to do it. This can only be done together, the same as building a new, better world.

# Additional Information

## *True Geometric Formulas*

| True value of pi | True value of e |
|---|---|
| $\pi = \dfrac{4}{\sqrt{\Phi}} = 3.1446055110\ldots$ | $e = [2\Phi^4]^{1/\Phi^2} = 2.718240808\ldots$ |

| Circumference | Area of a circle |
|---|---|
| $L = \dfrac{4D}{\sqrt{\Phi}}$ | $S = \dfrac{D^2}{\sqrt{\Phi}}$ |
| **Area of a sphere** | **Volume of a sphere** |
| $S = \dfrac{4D^2}{\sqrt{\Phi}}$ | $V = \dfrac{2D^3}{3\sqrt{\Phi}}$ |

| Squaring the circumference | Squaring the circle |
|---|---|
| $\dfrac{D}{L} = \sqrt{\Phi}$ | $\dfrac{D^2}{L^2} = \sqrt{\Phi}$ |
| **Squaring the sphere** | **Cubing the sphere** |
| $\dfrac{D^2}{L^2} = \dfrac{3}{2}\sqrt{\Phi}$ | $\dfrac{D^3}{L^3} = \dfrac{3}{2}\sqrt{\Phi}$ |

## For Chapter 6. Tonal Ratios within the Sequence of the Divine Spheres

In the cells of Table 1 below, you can find any frequency you need, and even the entire harmonious frequency sequence of square roots, if you know the vibrational frequency of at least one of the tones of this sequence.

| M | 1 | 2 | 3 | 4 | $F(N_i) / F(N_{i-1})$ |
|---|---|---|---|---|---|
| **N** | Layer 1 | Layer 2 | | | |
| **1** | **1** | 1.4142 | 1.7321 | 2 | |
| **2** | 0.7071 | **1** | 1.2247 | 1.4142 | 1.4142 |
| **3** | 0.5774 | 0.8165 | **1** | 1.1547 | 1.2247 |
| **4** | 0.5 | 0.7071 | 0.8660 | **1** | 1.1547 |

| $F(M_k) / F(M_{k-1})$ | 1.4142 | 1.2247 | 1.1547 | | $( F_0=1 )$ |
|---|---|---|---|---|---|

Table 1. This table allows you to get all the possible frequencies of tones of the "divine" sequence if you know at least one frequency.

The table columns M match these tones. The rows N correspond to the number of the known tone in the sequence. By placing this known frequency of tone in the corresponding cell highlighted by a bold contour, you can obtain the entire sequence for this frequency. The numbers in the cells of the bottom row and the far right column show the ratio of adjacent tones—that is, how many times the frequency of the corresponding tone is different from the frequency of the previous tone.

The first row of numbers in this table corresponds to a case where the base (lowest) frequency of the entire harmonious sequence is set. The base frequency can be any value, but, in this example, it is equal to one, so you can immediately see the ratios between the frequencies of neighboring levels.

The table has M columns and N rows, where both M and N are whole numbers from one to infinity. This table can be continued indefinitely, but, in this case, it is limited by two layers corresponding to the first four tones of the harmonious sequence. Dashed lines mark the boundaries of the layers. In this table, the first line is the main one. Suppose you place the base frequency value of $F_0$ (which is the lowest or the initial frequency of the sequence) in its first cell. In that case, you will immediately obtain the entire sequence because the values of frequencies in subsequent cells will be determined by the formula: $F_M = F_0\sqrt{M}$—where M is the column number or tone number.

If you want the second tone to sound with a set known frequency instead of the first one and want to know the rest of the tones of this sequence, you should go to the second row of the table. You can see that 1 moved to the second cell (the second column). If you know the third tone and want to know what the other tones of the sequence will be, you should go to the third row and so on.

It is easy to deduce the universal formula by which frequencies are generated in the cells of the table. If, let's say, we want to know what frequency the cell in the M column and the N row corresponds to, it will be equal to the following: $F_{MN} = F_{MM}\sqrt{M}/\sqrt{N}$, where $F_{MM}$ is your original known frequency placed in M row and M column, tied to your original sequence—that is, where N = M.

This table is very handy. It allows you not only to obtain a harmonious sequence of frequencies but also to identify other important ratios between tones.

If you pay attention to the bottom row and the right column, which cells contain the ratios between neighboring tone frequencies, you will see that these ratios are the same for both rows and columns. This is an important point when you consider that tones placed in rows form major combinations, and tones placed in columns form minor combinations. If you look at the ratios generating chromatic half-tones in the "traditional" musical scale, these ratios are 135/128 or 25/24, depending on whether this ratio splits major or minor tones. If we convert these proportions into decimals, they will be equal to 1.0546875 and 1.04167, respectively. If you extend this table, then you will be able to find more accurate numbers for these ratios both in our additional column and our additional row. Oher approximate proportions of whole numbers, which are known in music and have been obtained experimentally, are only an approximation of the exact ratios we have obtained in this table.

Table 2 provides the exact same table, but it is recalculated for the "special" frequency of 432 hertz.

| M | 1 | 2 | 3 | 4 | 5 | ... |
|---|---|---|---|---|---|-----|
| N | Layer 1 | Layer 2 | | | Layer 3 | ... |
| 1 | **432** | 611 | 748 | 864 | 966 | ... |
| 2 | 306 | **432** | 529 | 611 | 683 | ... |
| 3 | 249 | 353 | **432** | 499 | 558 | ... |
| 4 | 216 | 305 | 374 | **432** | 483 | ... |
| 5 | 193 | 273 | 335 | 386 | **432** | ... |
| ... | ... | ... | ... | ... | ... | ... |

Table 2. The same Table 1, but recalculated for the "special" frequency of 432 hertz.

Another small study I conducted revealed interesting laws of harmonic sequences. After all, you can make a similar table for any tone from this table, where it will be basic. If you start building the same tables for tones taken from the first, main, table, you will obtain many subsets of frequencies. When you combine these subsets into one table, you will get a multidimensional table of harmonic sequence. In this table, cells contain spectral bundles along with repetitive frequencies, when, for example, some intermediate frequencies appear between the neighboring main tones. In other words, the columns and rows of the main table can "exfoliate" — additional intermediate columns and rows may appear between them. This pattern allows you to understand well the depth of relations between all things. For example, if each person has a unique base frequency typical of this person only, one simultaneously has an entire sequence of derivative frequencies corresponding to each of one's organs. This spectrum forms a single harmonious sequence.

### *For Chapter 10. Some Unique Ratios of the Space of "Phi"*

While building this space, I found another curious ratio or number that I have called "Phantom Φ" (denoted by the letter Ψ), which is not as perfect as Φ, but in some aspects behaves similarly. Having discovered this strange number, I initially thought it was most likely Φ, transformed by the influence of a square root sequence. Look at Fig. 59.

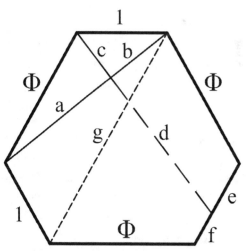

Fig. 56. One of the specific shapes where the ratios of square roots and the golden ratio are interconnected.

This symmetrical polygon consists of alternating segments equal to 1 and Φ. This shape is curious itself. If you connect its two vertices through one with a line segment $(a + b)$, this segment will be equal to the square root of two. Three of these connecting line segments form an equilateral triangle inscribed in this shape. The segment $g$, which is parallel to the side equal to Φ, connects two opposite vertices of this shape and is equal to $Φ^2$. Thus, if you draw a perpendicular $c$ from the vertex of the shape to the corresponding diagonal $(a + b)$, it will divide it into segments $a$ and $b$, and the ratio will result in the "Phantom Φ":

$$\Psi = \frac{a+b}{a} = \frac{\sqrt{2}}{a} = \frac{4}{\Phi^2} = 1.527864\ldots$$

This ratio is curious, because:

$$\frac{b}{a} = 0.527864\ldots$$

$$\left(\frac{b}{a}\right)^2 = 0.27864\ldots$$

$$\left(\frac{b}{a}\right)^4 = 0.07864\ldots$$

In all these ratios, there is the same "magical" endless fractional "tail" shortening each time by one digit.

If we extend this perpendicular until it crosses the opposite unit side of the polygon $(f + e)$, what proportion do you think it must have to split this side? Of course, it must have the Φ proportion. The ratios we have obtained from the analysis of this shape are very specific and revealing. Let's look at them all at once. So, in this shape:

$$g = \Phi^2; a+b = \sqrt{2}; f+e = \sqrt{1}; \frac{e}{f} = \Phi; d = 4c; d+c = \sqrt{6}$$

This example illustrates well how Φ "penetrates" our world of square roots. Fig. 60 and Fig. 61 show several more specific ratios and angles of golden ratio spheres. Similar magnificent ratios occur in pentagon space everywhere, which suggests that it is much more complex and perfect than the space of square roots.

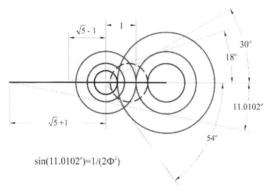

$$\sin(11.0102°) = 1/(2\Phi^2)$$

Fig. 57. Example of some specific angles and ratios for a sequence of multilayered spheres of the golden ratio.

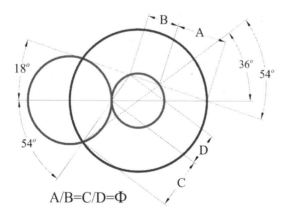

A/B=C/D=Φ

Fig. 58. Other example of specific angles and ratios for a sequence of the golden ratio.

If you try to build various combinations of spheres with diameters that are multiples of Φ, you will see that this can be done by an infinite variety of methods: spheres perfectly combine and approach each other, creating perfect ornaments. Therefore, these spheres can be studied endlessly because they are continually generating more and more new patterns and amazing spatial structures and ornaments.

To better understand the structure and harmony of the space of multilayered pentagonal spheres, let's first place a "golden ratio" sequence of spheres along one horizontal axis as if we have made a section of space and gone from layer to layer: from a layer with smaller diameter spheres to a layer with larger diameter ones, right to left (remember Fig. 44). If we portray them in this way, we will have a consistent sequence of spheres, with the diameters ever-increasing by Φ times and distances between tangential spheres ever-increasing by Φ times.

Let us remember now that our layers are in contact with one another by vertices of the icosahedral caps, resulting in spheres with smaller diameters being embedded in spheres with larger diameters. However, the picture is actually a little more complicated. Spheres are still arranged in such a way that they are never alone. If there is a sphere of some diameter, it should be tangential to exactly the same spheres while building its own space. Do you remember how we learned before that two identical spheres joining together result in a new independent object called "two spheres"? This new property refers to the same kind.

If there is a sphere with a smaller diameter in space, and at some distance (strictly defined by the size of the icosahedral cap) from it there is a sphere with a larger diameter, then there is also a sphere whose center coincides with the center of the larger sphere, and its surface is in contact with the surface of the smaller one. The sequence of spheres superimposed on each other result in a common perfectly harmonious picture of the tangential spheres embedded in each other, like a matryoshka doll.

### For Chapter 11. Obtaining Accurate Geometric Formulas

To obtain a squaring of the circumference (when the length of a circumference with the diameter D is equal to the perimeter of a square with the side L), the following condition must be fulfilled:

$$\frac{D}{L} = \sqrt{\Phi}$$

Let's see what would happen with the ratio of the circle diameter to the square side length at the true value of $\pi$ in the case of squaring of the circle, when the area of a square with the side of L is equal to the area of a circle with the diameter of D. It turned out to be an equally curious ratio:

$$\frac{D^2}{L^2} = \sqrt{\Phi}$$

It is also curious that for "squaring of the sphere" (the name of this term has come by itself, by analogy), when the area of the sphere's surface with a diameter of D is equal to the area of the cube's surface with an edge length of L, then,

$$\frac{D^2}{L^2} = \frac{3}{2}\sqrt{\Phi}$$

For the "cubing of the sphere," when the volume of the sphere with a diameter of D is equal to the volume of the cube with an edge length of L, then,

$$\frac{D^3}{L^3} = \frac{3}{2}\sqrt{\Phi}$$

The mathematics of spheres is perfectly accurate, concise, and different from the one we have been taught in school. However, that's not all. Do you know, for example, by how many times the perimeter of a square, circumscribed around a circle, is bigger than its circumference? Try to guess. It's bigger by exactly √Φ times! By how many times is the length of a circumference circumscribed around a square larger than its perimeter? It is larger by the square root of the square root of "Phantom Φ" times:

$$\sqrt{\sqrt{\Psi}} = 1.1178594\ldots$$

This ratio is of interest because, in the case of squaring the circle, the ratio of the circle's diameter to the square's side is described by a similar formula:

$$\frac{D_S}{L_S} = \sqrt{\sqrt{\Phi}} = 1.127838\ldots$$

However, that's not all! If we combine both kinds of squaring (a circumference and a circle) in such a way that, in the first case, we will impose two corresponding squares on the circumference, and in the second case we will impose two corresponding circumferences on the square, then the ratio (in the first case we mean the squares' sides, and in the second case we mean the circles' diameters) will be equal to the same proportion — that is, the square root of the square root of Φ. Thus, the square root of the "divine root" is also a remarkable value.

Now we know the basic ratios between squares and circumferences (circles). Let us also remember that if we extend the sequence of inscribed and circumscribed circles and squares so that the square circumscribed around one circle is inscribed in another one, we will obtain a sequence of circles and squares where diameters of circles correlate with each other as the square roots of whole numbers:

$$\frac{D_M}{L_N} = \frac{\sqrt{M}}{\sqrt{N}}; M, N = 1, 2, 3, 4, \ldots$$

Here, M and N are just numbers of circumferences within a sequence. The same applies to the ratio of square side length. That is,

$$\frac{L_K}{L_P} = \frac{\sqrt{K}}{\sqrt{P}}; K, P = 1, 2, 3, 4, \ldots$$

Here, K and P are numbers of squares within the same sequence. Areas of squares, as well as areas of circles, relate to each other like whole numbers.

In the end, let's remember that a square and a cube are not optimal figures for measuring area and volume. The optimal figures are an equilateral triangle and a tetrahedron. If instead of a square you put an equilateral triangle on a circle, and instead of a cube you put a tetrahedron on a sphere, and do calculations like those we did above, you get also something very interesting. Let it be your own discovery.

# References

[1]  Drunvalo Melchizedek, *The Ancient Secret of the Flower of Life*. Volumes 1, 2. Light Technology Publishing. 1998.

[2]  http://wiki.naturalphilosophy.org/index.php?title=Nassim_Haramein.

[3]  L. Horowitz, *Musical Cult Control: The Rockefeller Foundation's War on Consciousness Through the Imposition of A = 440 Hz Standard Tuning*. 2010. https://www.bibliotecapleyades.net/ciencia/ciencia_consciousscience26.htm.

[4]  N. Levashov, *The Final Appeal to Mankind*. II Publishing, San Francisco. 2000, http://levashov.info/English/books-eng.html.

[5]  T. Freke, P. Gandy, *The Jesus Mysteries: Was the "Original Jesus" a Pagan God?* Three River Press. 2000.

# O-BOOKS

# SPIRITUALITY

O is a symbol of the world, of oneness and unity; this eye represents knowledge and insight. We publish titles on general spirituality and living a spiritual life. We aim to inform and help you on your own journey in this life.

If you have enjoyed this book, why not tell other readers by posting a review on your preferred book site?

## Recent bestsellers from O-Books are:

### Heart of Tantric Sex
Diana Richardson
Revealing Eastern secrets of deep love and intimacy to Western couples.
Paperback: 978-1-90381-637-0 ebook: 978-1-84694-637-0

### Crystal Prescriptions
The A-Z guide to over 1,200 symptoms and their healing crystals
Judy Hall
The first in the popular series of eight books, this handy little guide is packed as tight as a pill-bottle with crystal remedies for ailments.
Paperback: 978-1-90504-740-6 ebook: 978-1-84694-629-5

## Take Me To Truth

Undoing the Ego

Nouk Sanchez, Tomas Vieira

The best-selling step-by-step book on shedding the Ego, using the teachings of *A Course In Miracles*.

Paperback: 978-1-84694-050-7 ebook: 978-1-84694-654-7

## The 7 Myths about Love...Actually!

The Journey from your HEAD to the HEART of your SOUL

Mike George

Smashes all the myths about LOVE.

Paperback: 978-1-84694-288-4 ebook: 978-1-84694-682-0

## The Holy Spirit's Interpretation of the New Testament

A Course in Understanding and Acceptance

Regina Dawn Akers

Following on from the strength of *A Course In Miracles*, NTI teaches us how to experience the love and oneness of God.

Paperback: 978-1-84694-085-9 ebook: 978-1-78099-083-5

## The Message of A Course In Miracles

A translation of the Text in plain language

Elizabeth A. Cronkhite

A translation of *A Course In Miracles* into plain, everyday language for anyone seeking inner peace. The companion volume, *Practicing A Course In Miracles*, offers practical lessons and mentoring.

Paperback: 978-1-84694-319-5 ebook: 978-1-84694-642-4

## Your Simple Path
Find Happiness in every step
Ian Tucker
A guide to helping us reconnect with what is really important in
our lives.
Paperback: 978-1-78279-349-6 ebook: 978-1-78279-348-9

## 365 Days of Wisdom
Daily Messages To Inspire You Through The Year
Dadi Janki
Daily messages which cool the mind, warm the heart and guide
you along your journey.
Paperback: 978-1-84694-863-3 ebook: 978-1-84694-864-0

## Body of Wisdom
Women's Spiritual Power and How it Serves
Hilary Hart
Bringing together the dreams and experiences of women across
the world with today's most visionary spiritual teachers.
Paperback: 978-1-78099-696-7 ebook: 978-1-78099-695-0

## Dying to Be Free
From Enforced Secrecy to Near Death to True Transformation
Hannah Robinson
After an unexpected accident and near-death experience, Hannah
Robinson found herself radically transforming her life, while a
remarkable new insight altered her relationship with her father, a
practising Catholic priest.
Paperback: 978-1-78535-254-6 ebook: 978-1-78535-255-3

## The Ecology of the Soul
A Manual of Peace, Power and Personal Growth for Real People
in the Real World
Aidan Walker
Balance your own inner Ecology of the Soul to regain your
natural state of peace, power and wellbeing.
Paperback: 978-1-78279-850-7 ebook: 978-1-78279-849-1


## Not I, Not other than I
The Life and Teachings of Russel Williams
Steve Taylor, Russel Williams
The miraculous life and inspiring teachings of one of the World's
greatest living Sages.
Paperback: 978-1-78279-729-6 ebook: 978-1-78279-728-9


## On the Other Side of Love
A woman's unconventional journey towards wisdom
Muriel Maufroy
When life has lost all meaning, what do you do?
Paperback: 978-1-78535-281-2 ebook: 978-1-78535-282-9


## Practicing A Course In Miracles
A translation of the Workbook in plain language, with
mentor's notes
Elizabeth A. Cronkhite
The practical second and third volumes of the plain-language
*A Course In Miracles*.
Paperback: 978-1-84694-403-1 ebook: 978-1-78099-072-9

## Quantum Bliss

The Quantum Mechanics of Happiness, Abundance, and Health

George S. Mentz

*Quantum Bliss* is the breakthrough summary of success and spirituality secrets that customers have been waiting for.

Paperback: 978-1-78535-203-4 ebook: 978-1-78535-204-1

## The Upside Down Mountain

Mags MacKean

A must-read for anyone weary of chasing success and happiness – one woman's inspirational journey swapping the uphill slog for the downhill slope.

Paperback: 978-1-78535-171-6 ebook: 978-1-78535-172-3

## Your Personal Tuning Fork

The Endocrine System

Deborah Bates

Discover your body's health secret, the endocrine system, and 'twang' your way to sustainable health!

Paperback: 978-1-84694-503-8 ebook: 978-1-78099-697-4

Readers of ebooks can buy or view any of these bestsellers by clicking on the live link in the title. Most titles are published in paperback and as an ebook. Paperbacks are available in traditional bookshops. Both print and ebook formats are available online.

Find more titles and sign up to our readers' newsletter at http://www.johnhuntpublishing.com/mind-body-spirit

Follow us on Facebook at https://www.facebook.com/OBooks/ and Twitter at https://twitter.com/obooks